Praise for A Farmer's Love

"A *Farmers Love* is not so much about farming as it is a diary of one man's life experiences, which include his love of farming. A heartwarming story." —**Frederick Kirschenmann**, Ph.D., President of Kirschenmann Family Farms and international leader in sustainable agriculture

"Walter Moora's story tells of his lifelong struggle to honor himself, his purpose, those he loves, and the land that holds us all is the overarching metaphor of our times as we work our way back to our place in the web of nature and forward into a new, more evolved way of being." —**Penny Kelly**, author of *Robes: A Book of Coming Changes*

"People everywhere are recognizing that one of the most important actions they can take for the health of their families, the environment and the future is to grow healthy food. In *A Farmer's Love*, Walter Moora offers inspiration and practical ideas for starting or deepening your journey for growing delicious, healthy food that will nurture you and a better world." —**Alisa Gravitz**, Executive Director, Green America

"Walter Moora...introduced us to biodynamics on his Wisconsin dairy farm several years ago. He balances intelligence and intuition in imparting his considerable knowledge and wisdom, and it is a great pleasure to be able to learn even more from this trusted resource. We're confident that Walter's book will do the same for others." —**Karen Page** and **Andrew Dornenburg**, former wine columnists for *The Washington Post* and coauthors of *The Flavor Bible*, winner of the 2010 Nautilus Book Award and 2009 James Beard Foundation Book Award

A Farmer's Love

Living Biodynamics and the Meaning of Community

Walter Moora

PORTAL
BOOKS

2011

2011

PORTAL BOOKS

An imprint of SteinerBooks / Anthroposophic Press, Inc.
610 Main Street, Great Barrington, MA 01230
www.steinerbooks.org

LIBRARY OF CONGRESS CATALOGING-IN-PUBLICATION DATA

Moora, Walter.
 A farmer's love : living biodynamics and the meaning of
community / Walter Moora.
    p. cm.
Includes bibliographical references and index.
ISBN 978-0-9831984-1-3 (alk. paper)
1. Organic farming—Philosophy. 2. Organic farming—Anecdotes.
I. Title. II. Title: Living biodynamics and the meaning of
community.
S605.5.M66 2011
631.5'84—dc22

2011002677

ISBN-13: 978-0-9831984-1-3
ISBN-10: 0-9831984-1-1

Printed in the United States of America

CONTENTS

I wish to dedicate this book to Susan, my beloved wife who has supported me so strongly in my dream to write this book.

…also to my two grown children Eve and David, whom I love dearly and am deeply proud of

…and to my late wife Joan, who shared so much of my adult life. I know I was blessed with her help many a time while writing this book.

ACKNOWLEDGMENTS

I want to thank Sherry Wildfeuer for spending many hours editing my book. Then on one of my trips back to the United States, she made time to sit down with me and help me clarify some of my thoughts on what underlies biodynamic practices. I felt like I benefited from her many years of biodynamic work and her dedication to producing *Stella Natura Biodynamic Planting Calendar*.

I also need to thank my beloved wife Susan for her thankless task of correcting my grammar and at times helping me state things more clearly.

CHAPTER ONE

LETTING GO

By the time I turned seventeen, I knew I wanted to farm, and most of my life I have followed my dream. There were times when I had to do other things, but during those times I was never happy. Then, in December of 2006, when I turned fifty-seven, I decided it was time to stop. This decision was not easy, because I was ready to continue farming until the bank closed me down. I had been on my Wisconsin farm for nine years, and it was my last chance to make a go of it. I really needed to make this farm work, because my whole identity was tied up with having a model biodynamic farm. It was more than just the money; it was part of my being, to show that biodynamic farming really worked and to demonstrate the richness of the soil farmed this way.

My poor wife Susan was watching me go down; she even put in a large chunk of her retirement money to support me and the farm. I was still enjoying the work well enough, but not the way I used to. Getting up at 4.30 a.m. on a crystal-clear January morning in Wisconsin with the stars wheeling overhead is reward on its own. Nevertheless, the challenge was gone. Sitting on a tractor for eight hours straight, trying to get the corn in before the rain, now seemed work that is more appropriate for a younger man. I was still too stubborn to quit, so my body just decided to make the point for me. Every few months my back would go out...I mean, really *out*. It was so painful that I would lie down and not be able to move. Just peeing into a bottle took ten minutes, ten minutes of excruciating pain. After taking muscle relaxants and pain killers for two days, I would be able to hobble into the Osteopath's office. Meanwhile, my herdsman would get exhausted doing all the milking and other chores. I would have to start working before being properly healed and still in pain, and that is scary. My body was telling me it was time to quit, but my mind was not yet willing. Susan has a good friend Martine, who is also psychic, and she helped me see the daylight so I could move on with my life. She sensed my fear that my life would be over if I failed at farming.

Before moving to Wisconsin in 1998, I'd had a small farm in Upstate New York, where we grew vegetables and baked bread. To keep me busy during the winters, I built a wood-fired brick oven that could bake as much as three hundred loaves each day. I specialized in European-style sourdough breads—the best in town, so I did very well. However, I felt that I should be farming and that I had missed my vocation. Although I was doing well financially and had a wonderful wife and two great teenage kids, I secretly wished I would die. I thought that perhaps in my next incarnation I could do better and get on with whatever I was meant to do.

Martine had detected my death wish and suggested I reprogram it to become positive. So we came up with three sentences that I would repeat to myself on a daily basis: "I will be fine if I am not farming. My life's work will not be wasted if I do not farm. And I can still live in the country and be associated with farming."

View of Mandango Mountain from our first house in Ecuador

The last sentence made it possible for me to make the transition into the future. Six months later, I decided to sell my cows and machinery and give up my farm lease.

Susan was ready for a change, because all our lives we had both been working exhaustively hard. We started talking about a sabbatical to restore and redirect ourselves. We went into the unknowing, and serendipity played with us. We now live in our own house in the Highlands of Ecuador, with the most beautiful view one can imagine. We hope to bridge the chasm between North and South for the good of both...and for ourselves. We live just above Vilcabamba in southern Ecuador, at an altitude of about five thousand feet. It was only in the 1960s that a road was finally punched through the mountain pass, which made the valley accessible to vehicles. Today, there are thirty small hostels and hotels, but it has retained its charm, with a small square surrounded by local shops, restaurants, and the church.

I feel at home here. I can hardly believe it but I often spend hours on the patio, immersed in nature. For close to forty years, I had mostly been farming, and now I wanted to experience "being " instead of "doing." I set myself an agenda, so that each morning before breakfast I would meditate, spend time studying Spanish, and then turn to writing this book. If any time was left over, I would hike or spend more time on the patio. As I write, I am realizing that I may reach more people to explore the wonder of biodynamics with this book than I ever could through a "model biodynamic farm," and that gives me great joy.

A friend suggested that I am comfortable here because I was born on the Equator in Borneo, and now I am back, but on the opposite side of the world at 5,000 feet and with a new language to learn. I think there are additional reasons. We belonged to a Quaker meeting for many years, and part of our spiritual practice was to tread lightly on the Earth and to live at peace with the rest of the world. Back in the United States, it is hard not to be dragged into consumerism; here, living without much is a way of life.

We recently returned to Wisconsin for a couple of weeks and I leafed through some of the old clothes catalogs. I began to feel like pulling out my credit card and ordering some nice shirts for $55. However, I didn't need them, since I had already bought some really nice five-dollar shirts back in Ecuador. And then there is the Homeland Security thing...it always sounds as though there is an Agent Orange Alert going on in the US. It's hard being part of the fear mentality and knowing that many things we do in the United States contribute to fear.

As we go through life, we have three congruent paths. Our life's work, or vocation; the group of people with whom we live or have karma; and our own personal development. These aspects come to the forefront at different times of our lives. Now I feel blessed that I can take time out and work on my own development for a while before "life" catches up with me again.

I have had two main themes in my life, both revolving around the care of the Earth. One has been to heal the Earth and produce good food through biodynamic farming, and the other has been to bring non-farmers onto the land. I want to teach everyone how we can care for the Earth in a spiritual fashion. I want to talk about the spirituality of the Earth from a farmer's perspective. I need some time to sink deeply into the beauty of the Earth and rejuvenate myself. With no conscious planning, life brought us here to one of the most beautiful spots on the Earth and we feel blessed. As for a particular country, I am rootless, having farmed on four continents, but I feel grounded in the world's earth.

Chapter Two

A BIRTH AND ALMOST A DEATH

When I was a young boy, my family moved a lot. My father was raised in Indonesia, a tropical haven, part of the Netherlands East Indian empire. My grandfather, being Dutch, and therefore a traveler, ran the post office in Surabiu, one of the big cities, and they enjoyed the comforts of colonial life. When he retired in the 1930s, my father was fourteen, and so they all moved back home to Holland. I think my father always longed to return to his boyhood home but World War II interfered, and then Indonesia demanded independence.

It was during the war when Germany was occupying Holland that my parents met and, despite the risks, married and started a family. My sister Liesbet was born in 1944, while Europe lay in ruins, and two years later my brother Johannes arrived. My father was eager to get away from the devastation of the war and back to his boyhood roots. He joined a company that was opening a lumber camp in the jungles of Borneo. This was in 1948, when Indonesia was recovering from the Japanese occupation and had just declared independence from Holland.

Getting to the village required a boat trip. I have no memories of that time, but think the trip took about five days after

SERIAL No. 008259

Name W. M. MOORA   Sex Male

Particulars of passport/travel document:
Type of document Dutch Pp.
Number E. 53-9563.
Date of issue 13. 3. 1957
Place of issue Singapore

2093—3,000—29-12-56—PT 100/20A—G.P., K.L.

My childhood passport

leaving Jakarta, Indonesia. The jungle where I was born is now a city of a half-million people, but at that time it was a native settlement with two or three other Dutch couples who were starting the lumbering. My father's job was to put in place the elements of infrastructure such as roads and electricity. Out of the swamps and jungle, my father planned and started a new city. I have never revisited that city, but once I saw it on the news because they were having riots there.

After arriving, my mother became pregnant with me. I don't think I was part of the plan, since my mother was trying to find her place in a small jungle settlement with her husband and two children. It was much too risky to have a baby in the jungle, where medical facilities were inadequate. Consequently, my mother had to take the boat back to Jakarta, where she and I could receive proper care. Unfortunately, she felt lonely in the strange city and nothing in her upbringing had prepared her for the situation. Being strong-willed, she promptly sailed back to be with her husband despite the risks. I was born with the help of the two Dutch women who were also living there. Apparently, there was a local doctor, but he had been sent away for being more nuisance then helpful. Nevertheless, he was able to save my life a few hours later. It seems I had contracted a tropical disease and developed a high fever, and there was nothing to do about it; I was going to die. I was in luck though. The doctor finally confided to my father that he had a new wonder drug called "penicillin." He knew only the dose for adults and hoped that he and my father could figure out a good quantity to give a twelve-hour-old baby.

As a result, I survived being somewhat unwanted and born into a hot, humid world with inadequate sanitation and a shot of penicillin. For a long time, I have had a feeling that my birth was not a happy time for me. Recently, I visited Jay, a hypnotherapist who helped me to reexperience my birth. As I entered the birth canal, I again felt the pain and anger of not being wanted. It was my right to have a loving mother to help me on my journey from spirit into the physical world, and I did not want to be born without that. It was my sister Liesbet who made me feel welcome in her love and enthusiasm for her new baby brother. Later, my mother was able to accept me and to shower me with her love and warmth, and I became part of the family constellation. My birth taught me to be independent and sensitive about being unwanted. It was hard then to be trustful of people. Looking back at my life, it is interesting to see how often I set myself up to confirm a belief that I am not wanted.

CHAPTER THREE

CAST OUT OF NATURE

After a challenging first day of my life, I believe life settled down but I have no way of knowing for sure, as my family history has pretty well disappeared. My parents both passed away some twenty years ago and Liesbet, who was five years older than me and who remembered more of our youth, died of a brain tumor more than two years ago. I do know that when I was about two, we had to flee Borneo.

At the time, Kelantan, or the bottom Dutch part of Borneo, became part of Indonesia, and the liberation troops came through firing guns. As they marched past, my parents made us lie on the floor with all the furniture and mattresses surrounding us. Later, we took the first boat out and landed in Malaysia, which was ruled by the British. For the colonists there, life was comfortable. We had a nice house with four servants, and the swimming club was five minutes down the road. Our mother would take us to the pool every afternoon, where we could play with our friends, staying cool in the water, while the ladies could sit and chatter or play Mahjong.

Actually, my mother got bored with this frivolous lifestyle and, since she was a trained kindergarten teacher, she started her own school. Young children loved her, and she had no problem handling twenty-five children with the help of the maid. It was not always so great for me. I did not appreciate sharing my mother with so many children. One day, I got so upset that I threw a temper tantrum, and my mother called the maid over and told her to take me to my bedroom until I behaved myself. This was a horrible experience for me, and I still remember the feeling of not being wanted and being banished for saying what I needed. I know that my mother had a job to do, but she could have handled it differently. I wished she had held me and let me know that I was the most important person in her life at that moment. That little episode certainly reinforced my feelings of not being able to trust the world, and that it was best not to show my feelings. Overall, though, I know I had a very loving home life.

Part of the tropical lifestyle was traveling back to the home country. Every three years, we would get a six-month paid vacation. Commercial air travel did not yet exist for most people, so we would take a three-week cruise from Singapore, past India, up through the Suez Canal, and on to England via the Mediterranean Sea. It was a wonderful, exciting trip, after which we would rent a cottage in Holland to use as a base as we traveled around Europe and visited family. On one of those trips, when I was seven, my parents had to decide what boarding school my older brother and sister should attend. In British colonial days, when children turned nine it was the custom to send them to boarding school to get a proper education and be indoctrinated into the system. They chose Michael

Hall School, about thirty miles south of London. It was a coed boarding school based on Rudolf Steiner's Waldorf education and allowed the three of us to be at the same school.

At the end of the vacation my siblings had to start school, so my mother and I stayed on in England for an extra six months. I was allowed to join class one, where I made many friends and loved my teacher. It was a wrench for me to go back to Malaysia and into the local army school. I still remember the day I decided that I'd had enough. I was a dreamy boy and had no idea what the teacher was talking about, which did not bother me. He called me to the front of the class and realized I was eating candy, which I did not know was forbidden. He promptly made me spit it out in front of everyone and stand in the corner. When I got home that day, I informed my parents that I would not be going back and that I wanted to go to school in England. I made such a fuss each day that they finally relented and put me on a flight to England with some friends. At the grand age of seven, I arrived back at boarding school during Easter, but was lucky that the three of us were all in the same hostel with about twenty other children. Liesbet would come and read me a story before bed every night and tuck me in. I loved school and still have some of the letters that my teacher sent to my parents. This one was written just after my return.

Dear Mrs. Moora,

    Walter seemed to settle in at once and seems very happy, except that he thinks he is too clever for Class

One and tells everyone so. I don't quite know why he thinks this, except he writes easily and well. However, he works hard in spite of this and enjoys the lessons. He is a strong character and good boy, who will lead the other children in the right way. I can imagine how you must miss the children. I think it was very brave of you to let Wally come, but I am sure he will be much better off away from that school.

With best wishes,
Joyce Russell.

Another letter that was part of my first class report went as follows.

Wally works well on the whole, though he is fond of play time, too. He doesn't say that he is too clever for the class anymore, as he realizes now that some of the others know more than he does. He is just about right for his age, as he knows the letters and will be ready to start reading next term. His writing is very good and his number work about average. His chief friend is David Newbatt, as they are together in the hostel and both are choleric and can stand up to each other. They are very good for each other. Wally is friends with all the children, he has been asked out many times. He is so good-natured and such fun that they all love him.

Yours sincerely,
Joyce Russell.

I have very fond memories of that time but I always felt a little different, as I had a stranger's accent wherever I went. Even after living in the United States for more than thirty years, people still ask me where I come from. With her three children in England, my mother was torn between being with her husband in Malaysia and with us, so when I was nine we all moved to New Zealand so that we could be together as a family.

Our first house there was in the country, surrounded by fields and bush. As children, we were able to roam the countryside and explore nature. I remember one afternoon in particular, as I was passing through a patch of bush, that somehow the birds sounded different. New Zealand is famous for its many native birds that are well known for their beautiful songs, and I had always felt at one with them. But now the birds felt outside of me, as if I wasn't part of nature anymore. I was a spectator and felt horrible. It was a fleeting moment, but I have always remembered having this experience of being cast out of paradise. I have spent much of my life trying to regain oneness with nature and the spiritual world.

CHAPTER FOUR

THE FLIGHT OF A BIRD...
AND BIODYNAMICS

My life seems to go in cycles, and every seventh year things change for me. When I was seven, I chose to go to boarding school, and then again at fourteen I made a major change. I was very close to Liesbet, and it was then that she married a sheep farmer. Her husband-to-be Rod had a beautiful hill farm with seven thousand sheep and four hundred beef cattle, all on sixteen hundred acres. The farm looked like a park and was located at the end of a seven-mile, gravel road. The land was divided into many paddocks, and the greenness stretched from one hill to the next. In the hollows were lakes and ponds and patches of native bush. Here and there would be flocks of sheep and cattle, and a quietness pervaded the whole.

During my vacations, I would stay on the farm and help with the work. All the stock work was done on horseback, and at lambing time, we would spend six to eight hours riding around, helping any ewes that were having problems and saving lost lambs. It would be springtime and a beautiful day with a blue sky, the hawks wheeling overhead, the lambs playing and the gentle calling of the mothers. I felt that life was good. There were other days when the rain would drive across the saddle, and by lunch my hands would be so stiff and cold that I could barely undo the girth straps and take off the bridle. This, too, brings back memories of animals saved and good work done.

In contrast to this, school was not so fulfilling. Looking back, I realize that high school could not give me what I needed. I wanted to be part of the world and to succeed. I worked hard to be successful in academics and sports. I had friends, but by sixteen I felt disappointed in humanity and was pining for the solitude of the farm. Though I did not realize it at the time, school had no room for my search for spirit, and I learned to hide my true being. I liked farming, which allowed me to get away from people, and I really enjoyed the work and being in nature. Nature can seem hard and indifferent, but it accepted the way I was. I decided that my vocation would be farming.

I graduated from high school in 1966, when the hippie movement was in full swing and we thought we could change the world. Rachel Carson's book *Silent Spring* confirmed my suspicion that we were fighting nature and traveling down a destructive road. The photographs of our planet from outer space were particularly moving to me, letting me experience the beauty and fragility of the Earth.

After high school, I worked on farms for two years and then went to agricultural school for two six-month periods, with a six-month farm stint between them. During those first two years on farms, I was doing correspondence courses and after all this I got my agriculture diploma. This sequence gave me a good grounding in conventional farming while searching for a way that would acknowledge the spirituality of the Earth and allow me to work with these forces.

I was lucky, in that my parents were sympathetic to my interests and introduced me to biodynamic farming. Here, I found a belief that the physical world is a reflection of the spiritual world and that we can work directly with those spiritual forces. At that time, I was working on a beautiful farm on a plateau, wedged between the ocean and mountains. I spent many hours sitting on a tractor doing fieldwork. By modern standards, they were small tractors, with no cab, air-conditioning, or radio. I was not so cut off from nature as I would be today on a modern tractor.

Plowing was especially fun. It took real skill to keep the rows straight and even. As the front wheel of the tractor moved slowly down the furrow, I could look behind and watch the earth roll over in long ridges. The gulls would follow all day long, searching for grubs and worms, and it was very peaceful. I would be by myself all day and had no access to media such as newspapers or television.

I had become interested in Anthroposophy and would study during the evenings. One night, I was lying in bed and meditating when I experienced myself floating in the corner of the room, looking down at my body. This really freaked me out because I did not know how to get back in. The shock drew me back into my body, but because I had no one to talk to about the experience, I decided to stop that particular meditation. Now, of course, I would love to repeat the experience, but my mind is too full of distraction.

While working on that farm, I had an accident and cut my wrist with a chain saw. I cut two tendons, and at the hospital, my arm was put in a cast. I could not work for two months and took this opportunity to visit some biodynamic farms. I was especially taken by one of the farmers and how he stopped our conversation to look up into the sky to follow the beautiful flight of a hawk. He was getting close to retirement, and I sensed that he had a deep wisdom and gratitude toward nature. If this is what biodynamic farming does for a person, then I wanted to follow in his footsteps. Soon after, I left New Zealand to learn about biodynamic farming in England.

CHAPTER FIVE

FALLING IN LOVE

My brother Johannes was living in Northern Ireland an intentional community that included a biodynamic farm, and I decided to join him there. The transition from four years of isolated farms and agricultural school to a very intense community of two hundred people was difficult for me. Glencraig, as the community is called, is part of the worldwide Camphill Movement that works with people with developmental disabilities. It is a magical property with a small herd of dairy cows and extensive gardens and sits right on the Belfast Loch. The work on the farm was tailored to the needs of the developmentally disabled and we tried to keep the work place simple and people friendly. At that time, we milked six to eight cows by hand and there was always lots of shovel and wheelbarrow work for those who could do it. On the other hand, we did have tractors and equipment to do the heavy work such as haymaking.

Glencraig also had a school for children with developmental disabilities and a three-year training course for the coworkers. The way that Camphill works is that everybody is a volunteer with no pay, and we all shared our lives with the disabled. All our financial needs were taken care of, including vacations, and I felt good about this arrangement

At Glencraig, I experienced that care of the land is a community activity and not the responsibility of only the farmer. Each Sunday morning, we had a land group meeting attended by many of the community members. The farmers would give reports of the week's activity and future projects. Because the community attempted to be self-sufficient in food, the housemothers were always interested in the vegetables and other produce, including the number of gallons of milk they would receive each week, and sometimes the farmers would voice complaints. It happened one time that the cows got out because someone had left a gate open, and consequently the vegetables were trampled. We also discussed long-range needs and how they fit with the community resources. Questions always arose, such as whether we needed a new tractor and, if so, how we would pay for it.

What spoke to me most strongly at Glencraig was the spiritual foundation of the community. Every morning after breakfast, the households would gather and read a passage from the Bible. Then, on Saturday evening, the houses would attend a "Bible evening" based on the Last Supper. We would sit in silence for twenty minutes, and then a candle would be lit and we would move to the dining table to share a simple meal of bread and salt and a glass of grape juice. While eating, the conversation would revolve around various people's special experiences during the week while avoiding day-to-day troubles. Then the table would be cleared and the Bible read. The person reading would lead the conversation as everyone joined in. It was central to community cohesion to read the Bible each morning and, once a week, join in communion with the other members. We also celebrated the festivals with artistic activities such as dramatic productions based on the spiritual aspect of each festival. There were also opportunities to participate in study groups, during which one of Rudolf Steiner's books would be read. Often on Sunday evenings, one of the community leaders would give a lecture.

I loved the spiritual aspect of the community, but after six months of such intense living, I burned out. I still struggled with feelings of discomfort with people. I must have been shy, because people might joke, "Walter said three words today!" From my perspective, I was lying low and seeing how everything worked after having been alone on farms for so long. For one thing, there were more young women than men, and I knew I was being checked out.

The previous three years had been lonely for me, and I was a little overwhelmed by all the people and community activities. In the evenings, the young coworkers were expected to lead activities with the handicapped adults, and I found myself supervising sessions at the heated swimming pool with one of the women coworkers. Joan wore a bikini and had shinning dark eyes, and I fell for her. She lived two houses down the drive, and I noticed that around two o'clock every Sunday, she posted her letters home. I would give her a few minutes head start and then saunter out and accidently bump into her. She was much more astute then me. Years later, she told me that the first time she saw me she knew we would get married.

Part of my troubles at Glencraig stemmed from the fact that I was exhausted. Being young and in love, Joan and I would chat and make out until midnight or one in the morning before going to sleep. It was summertime, and being so far north, it starts to get light soon after 3 a.m. in Northern Ireland. I would wake up and wait for six to roll around, when I would go milk the cows. Paul, the other farmer, was married and had a baby that cried all night, so he didn't come to milk on a regular basis. Milking eight cows by hand by myself before breakfast with too little sleep was not the right way to start the day. After six months, I needed a change, and I applied to Emerson College, an anthroposophic school in Sussex, England, where I was accepted.

It was wonderful to be a student with little responsibility. After Glencraig, where everything was so intense, I needed to stand back and absorb everything I had experienced. I missed Joan terribly; she was upset with me for leaving Glencraig and doing my own thing. She had decided to return to the US and get her master's degree in Waldorf education.

At Emerson, I could be as engaged with the other students as much or as little as I wanted. The school is based on the work of Rudolf Steiner and offered a foundation year of anthroposophic studies, which could lead into a second year of specialization. No credits or grades were offered so it was totally a self-enrichment program. The mornings were more intellectual, with lectures and study groups, and the rest of the day would involve activities like arts and crafts and woodwork. If life got too intense, I had my favorite spot in the library where I could read or I would play hooky and go for three-hour walks in the Ashdown Forest.

Meanwhile, back in the States, Joan decided that she would take a teaching position at the Kimberton Waldorf School in southeastern Pennsylvania. This was tough for me, since I had no interest in moving to the US. However, Joan's parents offered me a job painting their house, so I spent the summer painting on Long Island. By the end of the summer, we decided we would get married and move to Pennsylvania.

A five-hundred-acre farm was attached to the Waldorf school, but it was still being farmed conventionally by the old farm manager. Across the river, a new Camphill community with 350 acres of land had just been started. The land was going to be farmed biodynamically, and I moved there to help in the farming operation. The land had been the estate of Alarik and Mabel Pew Myrin, who were associated with Sunoco Oil. Mr. Myrin was a student of Rudolf Steiner's work and had helped establish Kimberton Waldorf School in 1941. Following his death, the rest of the estate was given to the Camphill community.

I was twenty-four when we married. The wedding service was held in the old ballroom of the mansion. As the priest walked into the room, I experienced a moment of confusion, and the veils to the spiritual world fell away. The room was crowded with our friends and community members, while surrounding and filling the space above and around us spiritual beings looked on and celebrated our sacrament of marriage. I felt that both our friends and those spiritual beings witnessed our vows. It was a reminder that our world is not so separate

Walter and Joan with in-laws and Walter's sister and husband

from the spiritual world, and that our actions are important on many levels.

Joan and I spent our next four years in Camphill, Kimberton. She taught at the Kimberton Waldorf School and commuted back and forth. She really was an excellent teacher; she put her heart and soul into it and the students loved her. Being part of two intense communities was not easy, and we realized that eventually we would leave. Waldorf schools ask for a commitment of a certain time from the teachers as they move with their classes through the grades. After the end of fourth grade, we started looking for our own farm.

Meanwhile, I had been given a wonderful opportunity to take over management of the farm and to begin the conversion to biodynamic agricultural practices. I not only learned about running a farm, but also about living in an intentional community. We milked about forty cows and farmed three hundred acres. By the end, there were four coworkers and three mentally handicapped people on the farm. It seemed like too many people for me, but that is part of the work of Camphill and community living. In addition, I really wanted to find out if I could farm on my own without the support of a community.

When the school year finished, we packed up and started looking for our farm. We were both twenty-eight and ready to find out what we could do on our own.

CHAPTER SIX

CRESSET FARM

Joan wrote a book about our first nine months on Cresset farm, but it was pretty sanitized. For Joan, writing allowed her to leave the hard times behind and remember the good. She put her heart and soul into the farm and loved the land, but reading between the lines, it was rough. It was primitive; it was hard work; and all the money went into the farm. She had little knowledge of farming and it was hard for her to have much input, so she followed my ideas.

While we were in Camphill, we had saved $28,000, which was not enough to buy a farm, cows, and equipment. Our lending institution was Farmers Home Administration, which worked with us and gave us a $100,000 thirty-year mortgage at three percent on the farm and a $40,000 seven-year loan for our cattle and machinery. The monthly payments on those loans put a lot of pressure on our budget and left little margin for error.

We did have our 110 acres of good soil, but the barn was run down and the dwelling was a large old, uninsulated Upstate New York farmhouse. In the winter, it would get so cold in our kitchen that peas soaking overnight in preparation for pea soup would freeze. Looking back, I could never do that to my

*Joan on vacation
shortly after our marriage*

wife and children again, but it was also a wonderful time. I still feel that it was my best time, even if on some levels it failed.

As a young man of twenty-eight, it was wonderful to put my heart and soul into developing the farm. After two years, we were able to buy neighboring land with 120 acres. This made it possible to grow all our own feed. This was important to me as part of the biodynamic model. Soon after, I took a course in homestead cheese making, and we were able to build a cheese house, where we made Gouda cheese. We called our place Cresset Farm. A cresset is a vessel that holds precious oils, and we felt that the farm was like a container within which special things could happen.

Although Joan was raised in the suburbs of Long Island and was used to having people around and a high standard of living, she willingly followed me in my dream to be part of the back-to-the-earth movement and raise our family. She was not raised as a farm girl and never did learn how to do such things as driving a tractor, but she was certainly indispensable in other ways. I remember the day our neighbor drove into our yard and needed our tractor moved so that he could get out to the back.

He just assumed that Joan could move it, and a look of amazement came over his face when Joan informed him she didn't know how to start it.

Back in the 1980s, most wives on family farms could take their turn sitting on the tractor, doing chores such as raking the hay. But Joan did much more than that. She created the feeling around the house and yard that made it nice to be around. She created community for us. She had a social skill that made people feel at home and wanted. I have many special memories. In the summer, after a hard day of making hay, we would all go down to Long Point State Park on Cayuga Lake. At that time, it was still undeveloped and there would be few people. It was wonderful to relax, play with the children, swim in the cool water, and then have a picnic dinner and talk as the Sun went down. By then, we had Eve and Dave, and they would sit happily on our laps as the evening became quiet and dusk changed to night.

Joan, a born teacher, would tell endless stories to the children and knew all kinds of games that kept them busy on long winter days when it was too cold to go outside. In the summer, when there was fieldwork to do, David enjoyed coming out with me on the tractor. After lunch, Joan would happily give me David and promise to come get him soon. I knew that the tractor would lull David to sleep after ten minutes or so, but this did not bother Joan. He would be sitting in my lap, and after about ten minutes fall asleep. For the rest of the time I would have to hang onto him and prop his head up so it would not flop around. All the while, I would need one hand free to steer the tractor and lift and drop the implement at the end and beginning of each row.

Meanwhile Joan would be enjoying her free time from motherhood and delay her promised return. After lunch, it was nap time for the children, and often it was my job to make sure they didn't play while falling asleep. I would lie on the floor and sometimes I would fall asleep before the children did. They would notice and creep away, only to be caught by their mother, who would also scold me for failing my duty.

We celebrated festivals at special times and invited friends, or later, after Joan started teaching again, the families from her class would visit. People love to visit farms that are still on a human scale, and stanchion barns, where the cows are tied and handled daily, are cozy places to be. For children, it is especially nice when, at milking time, your teacher or her farmer husband can help you wash a cow's udder and then squeeze the milk out of the teat.

When we first started to make cheese, we had cheese festivals. One year, about seven hundred people visited on a single day, with cars parked up and down the road. We had a 7,000-pound cheese vat and usually made about 500 pounds of cheese every second day. For the festivals, we made cheese in the afternoon so that people could see the process, while outside clowns and musicians entertained. We kept the cows inside, and everything was spic-and-span. We also had two teams of horses that pulled hay wagons for people to ride on to the back of the farm.

The timing was just right. It was a beautiful fall day, with all the trees in full autumn color. Especially nice for us was watching Dave, who was four. He sat next to the driver, content as could be, for the duration of the festival. The festival was a financial success for us. The community discovered our cheese house, and we had many repeat customers over the years.

However, life was also difficult. We were cash-strapped, and it was hard to sell all our cheese. We had forty-five cows and enough milk to make 500 of cheese every other day. This was in the early 1980s, a time when there was a huge surplus of milk and milk companies were not taking on new customers. As a result, although we kept making cheese and filling our aging room, we had no money coming in. In addition, the value of the dollar was high, which meant I could not compete against imported Gouda cheeses. Imported cheese was coming in at $1.80 a pound, whereas I had to sell my cheese at $2.25 or sell the milk. It became impossible then to wholesale cheese without losing money.

Even worse, the New York Department of Agriculture and Markets was giving me a hard time. Farming is heavily regulated, and everything has to be inspected and approved, including labeling. Ag and Markets was telling me I could not call my cheese "organic," since all cheese is organic. This was very upsetting, because "organic" was what set my cheese apart from the competition. At the time, organic food was just coming into vogue, but official standards were not yet in place.

Officialdom can be intimidating so I went to my lawyer, who advised me that my farm was worth less than the cost of suing Ag and Markets, but that I should talk to my local assembly member Steve, who happened to be head of the

Assembly Ag committee. Steve knew that Ag and Markets had been asked two years previously to come up with organic standards but had not bothered. Steve kindly inquired where the standards were for organic cheese, so I could understand their ruling. One week later, I got a letter saying I could call my cheese "organic," but, in the future, please talk directly to Ag and Markets. Yes, it always feels good to beat the system.

It was difficult to depend always on interns to get the work done and share our home with them. It was especially hard in winter. Our sole source of warmth was our wood stove in the living room. The bedrooms upstairs could often get below freezing, so we crowded around the stove to keep warm. A family with two young children and a couple of interns in their twenties was not always compatible. Getting up at four-thirty in the morning to milk the cows was especially painful. I would get up without turning on the lights so as not to wake Joan. All the clothing was set out in a special order, long johns first, with the tops still inside the shirt and sweater all ready to be pulled over the head without twisting up and then the bottoms and jeans and socks. Dressing was very quick. Downstairs, the woodstove would need to be stoked and coveralls put on. The insulated boot liners were always left under the stove, so the feet were warm and dry. Once in the barn, it was warm, as the cows' bodies and breathing kept the barn above freezing.

I had much to learn about winters in Upstate New York. In the fall, when it was wet, we would make ruts in the lanes when we hauled the corn in from the fields or took out the manure. Then in the winter, the ruts would fill with water and freeze. During cold weather, the cows were in the barn most of the time, and the gutters behind the cows would have to be cleaned each day. Cows make enough manure every day to fill a manure spreader, and it has to be hauled out to the compost piles before it freezes. That first year there were many days when my tractor got stuck in the frozen ruts, and I had to go out with a pick and break up the ice. A couple of years later, I was able to buy a four-wheel-drive tractor, which solved that problem.

Eventually, my vision of why I was farming became lost with all the hard work and financial pressures. I was working eighty or ninety hours a week just to get the work done and forgot about the spiritual side of life. I was just slogging it out to prove I could make it. This was not enough support for Joan. In the Christian Community wedding ceremony, the priest turned to me and said,

> Walter, shine before Joan
> With the light
> Which the Risen One
> Lets shine in your spirit

He then turned to Joan and said,

> Joan, follow Walter
> In the light
> Which the Risen One
> Lets shine in your soul

It was Joan's feeling that I was not embodying the spirit light she needed to follow, and therefore she wanted to leave. This was devastating to me and I did not know what to do. I loved Joan and the children and could not imagine life without them. The farm would be empty without them, but I knew only how to farm and could not imagine providing the family with what they needed without a farm. I tried to be supportive of her needs. Her parents visited, and they went apartment hunting with Joan in Ithaca. Joan applied for Waldorf teaching positions on the East Coast.

I remember one night in February, leaving Joan in Boston for a job interview. I was driving through snowstorms with Eve and Dave in the back. We could not leave Boston until six that evening, when the snowplows had finally cleared the roads. I had promised my intern that I would try to get back to the farm for the morning milking. It was still snowing a bit, and I drove all night on the Massachusetts Turnpike and New York Thruway, wedged between tractor-trailer trucks and the snow coming at the windshield. We tried stopping for a rest, but it was too cold and we could not afford a motel.

Years later, Eve told me she never fell asleep that night; all they knew was that Mom and Dad were changing their lives. We eventually arrived home in time for me to put my children to bed and go out to milk. I struggled on while Joan tried to find the right place for her and the children. After I wrote this chapter, I asked Eve if she had any memories of those times and, as a thirty-year-old women, she wrote back:

Hi Pops,

It is so nice to read your story and bring back memories from that time. Your writing is beautiful. I have many memories from our time at Cresset farm. One of my best memories is of Christmas on the farm. I remember going out with you and Dave to pick out and cut down a Christmas tree on Christmas Eve, and then hanging popcorn strings and apples on the tree with mom, along with our other ornaments (those gold cymbals and candles). I remember how we went out to the barn and sang to the cows, and I remember singing around the tree and telling stories during the holy nights. I remember Christmas quite well in that old house. I remember the fire being lit in the stove near the tree.

I also remember the festivals on the farm and how fun it was to have all those people come visit the farm. I remember times when we had babysitters, and I was always so sad and upset when you left (when I was real young). Many of my memories are actually around traumatic experiences, like when mom got in that car accident and red paint spilt on me and scared her. I remember playing with Laurie and Jordan and Russell, too (and I remember when he died). I remember the auction of the farm and that mom had a splitting headache when it was over and was lying on the couch upstairs in the addition that we added onto the house. I have a memory of seeing a ghost (or some spirit) walking thru my room one early morning and I remember our intern Daren. I could go on with lots more memories, but I'll talk to you about them if you want.

Anyway, loved reading your chapter.

Around this time of great sadness and uncertainty, I received a great gift that sustained me and changed my life. I had just gone to bed and Joan was still downstairs, when a mighty being entered the room. This being radiated light and love and communicated to me that I was completely loved and accepted just the way I was. I knew that I had an unbelievable companion that would never leave or stop believing in my goodness. After a while, the being was gone and I was left in tears of gratitude. I knew that whatever happened and however painful, things would be all right. Somehow, this experience changed me so that I could again be the light that Joan needed to experience shining before her. I do not think much happened outwardly, and I still had to work incredibly hard to keep the farm going. Yet I knew that all creation was perfect and loved to an extent unimaginable to us.

Ever since we moved to our farm, Joan had been helping to start a Waldorf school in Ithaca, about thirty miles away. When Dave turned three, she started to teach the kindergarten class. The commute was long, but she loved the teaching and meeting more people. Then the school had to change location, and the commute became more than an hour each way. Joan and the children had long days and would come home tired. We were still struggling with finances on the farm, and I felt we were spinning our wheels and going nowhere. With all the hard physical work, my sciatic nerve was bothering me and causing me to limp.

Finally, after seven years, we decided to sell the farm. At that time, in 1985, the dairy industry was in disarray and our two neighbors were also going through bankruptcy and having auctions. The investment in our cheese house had no resale value, and with the depressed value of dairy farms, the Farmers Home Administration said they would buy back all our land and assets for one dollar and forgive all our loans. We were allowed to keep our car and about ten thousand dollars worth of cheese that we sold over the next few months. I was able to sell all the cheese equipment, which I then had to deliver to Wisconsin.

I still remember shutting the door to the U-Haul truck before driving out, thinking that I was closing the door to all my dreams. It is strange how I thought I was selling the farm for the good of the family, but years later Joan said she loved it there and would have stayed.

CHAPTER SEVEN

SEARCHING FOR OUR PLACE

We were a family of four. The year was 1985. Dave, the youngest, was just turning five, and Eve was ready for first grade. Joan and I were both thirty-five. To keep my mind off the farm and the auction, I had promised us a trip around the world, which would include New Zealand. Losing a farm is gut-wrenching and a bit like losing a child. I knew I would not be able to process it right away, so it was better to put my mind on something else for a while...like a trip.

A farmer is defined by his or her place and work. Although our farm was still run down, people who knew about farming said we had really cleaned the place up. I had been making cheese I was proud of and was selling it around the country. In my work, I had a daily rhythm of milking and chores and the yearly rhythm of the seasons. Always rushed in the spring, trying to get the crops into the ground, the summer was spent making hay and the fall getting the corn in or filling the silo one last time. Even winter was busy with feeding and keeping the cows clean and comfortable. There is not much time to think about who you are. I was a farmer with a family. What else was there than that?

After selling the farm, I had to confront that existential question and try to work out who I am without a farm. Farming has two sides to it. On the one hand, it can be very spiritual. It is all about birth and death; it is about creating and destroying matter. In the sacrament of communion in the Christian church, the bread and wine are raised symbolically to the flesh and blood of the risen Christ. Farmers who spend their whole day working in the physical world are like priests working in nature. We are constantly transforming matter into living substance that feeds human beings. On the other hand, farming is very physical, and it can drag you down into materialism. Running a farm is a bit like running a truck business; you have to haul feed to the cows and then take all the manure back out to the fields. It pays very poorly and involves long hours. Maybe that's why it ranks last on many young people's choice of professions.

I was somewhat lost about our future, so Joan and I stored all our things in a friend's barn and left. This time, we went to look around the world for our place, not just around the East Coast. We decided that we needed to retrace our earlier lives and see if we could fit into one of those places again. After all the hard work, it was nice to relax, and my sciatic leg soon got better.

We were still interested in the Camphill movement and eventually joined Mourne Grange, a Camphill community in Northern Ireland, not far from where Joan and I had met. I ran the small dairy farm, and Joan taught the children. However, we learned that it was not the right place for us. Although I truly admire the intentional communities of the Camphill movement,

I was not comfortable and wanted greater independence. I felt that by joining Camphill I would have to immerse myself in the community and give up my individuality. Later, I felt, I would rise again, remade and stronger, but I was not yet ready to do that at the time.

After six months at Mourne Grange, we went to New Zealand, flying Singapore Airlines. The airline had given us a special deal, whereby we could have a three-day stay-over at almost no additional charge. A nice hotel and all kinds of tours were included. Our little family had a wonderful time being tourists. Singapore was celebrating the New Year, which made it a little noisy and crowded, but we went to temples, traditional dance shows, and the usual places to buy stuff. This was very different and exciting for our country children.

We stayed in New Zealand for two wonderful months. We stayed at my sister Liesbet's farm, which had a beautiful kiwi-fruit orchard right on the river. My parents lived only twenty minutes away, so Eve and Dave finally got to know their grand-parents, and I was able to reconnect to my roots.

While we were there, my brother Johannes visited with his wife and two children. We had decided that we would rent a camper van and spend three weeks touring the South Island. The four children were new to each other and loved playing together, and the wives also became good friends. It was more a question of whether the two brothers could survive each other in a cramped space.

Everything was fine; we took turns driving, until I drove under a low overhead bridge and demolished the top of the camper. Luckily, we were close to a distribution point, and another van would arrive in three days. I have to admit I was a little shook up, but Johannes and I decided that we would not rent a car for the next three days, since New Zealand has such a good bus service.

Lo and behold, ten minutes later, as we were unloading the camper van and without any further discussion, Johannes informed me that he had decided to rent a car. This brought up all kinds of boyhood memories—mainly of me tagging along as the younger brother and my older brother walking all over me. I had spent the last twenty years getting over this, and I was not about to fall into old patterns.

As I came down the camper van steps with an armload of toilet paper, he informed me of his decision. I started yelling at him as I threw rolls of toilet paper in his direction. He drove back to our friend's house and I caught a bus and walked. We continued our verbal fight in front of the kids and wives, with me threatening to end the vacation. However, our wives managing to smooth things over.

My brother's parting words were that it was my problem how I felt about our boyhood and that it was not his fault. He was sorry that I was upset, but not sorry about how things were when we grew up. Maybe he had a point that it was my problem and not his. I felt good that I had stood up for myself instead of allowing our relationship to fall into old patterns.

Recently, Johannes and his partner visited, and in conversations about our youth he did say that he was sorry that it had been so hard for me; he had not been aware of my situation.

This made a huge difference to me. The way we grow up in a family sets our patterns for life, patterns that are hard to break out of. Sincere apologies can transform a situation into a positive experience. We ended up having a wonderful vacation, and at night Joan and I got to sleep in the tent, rather than in the crowded camper van with all the children.

While in New Zealand, we visited several biodynamic farms but did not feel called to stay there. At the end of two months, it was time to find a place to be, and I accepted a position on the farm that belonged to the Kimberton Waldorf School, where Joan had taught eight years previously.

CHAPTER EIGHT

WASHING WINDOWS, GROWING VEGETABLES, BAKING BREAD, AND BACK TO FARMING

The farm that belonged to the Waldorf School had been bought by Mr. Alarik Myrin back in the 1930s. He had hoped that it would become a model biodynamic farm and school, but it didn't work out that way. A Waldorf school was started on the land, but the farm had been run using conventional farming practices for many years. A few years before my arrival, management had changed and a group of young farmers were running the farm biodynamically. The situation was perfect for us. There was a Waldorf school for the children, and in the future Joan would be able to teach when a position became available. There was also a school farm program in place, and classes would come to the barn to help with chores.

The farm had many beautiful places to walk, so we invited the school families to visit whenever they wanted. We also created some wonderful festivals. In the spring, we had a farm blessing based on Rogan's Day. This was the day the early settlers blessed their land and asked for good crops during the coming year. We had the children pull an old horse-drawn, one-bottom

A class outing on our farm

plow through the garden with a long rope while we all sang. We then placed a loaf of bread made from last year's wheat in the furrow, covered it with soil, and asked for a good harvest that year. We finished the day with a hayride around the farm.

We were able to help start Community Supported Agriculture (CSA) on the farm. We integrated it into the farm organism by giving them cow manure and including the gardeners in some of the decision-making on the farm. The new farm store and the CSA, along with the animals and the beautiful land provided real opportunities to feel connected to the land for those members of the school community who wanted it.

When we arrived, the farm was showing a financial deficit, and the faculty and board were trying to find ways to eliminate the loss to the school. Construction had already begun for a

The class with our pig

farm store and a milk-bottling and yogurt plant. Because the farm was on Seven Stars Road, we changed the name to Seven Stars Farm, and the yogurt was sold as Seven Stars Yogurt, which is now marketed successfully nationwide.

I also put together a proposal that the farm business be run and owned separately from the school as a for-profit business. The school board thought it was a wonderful proposal, but the other farmer was against the idea of separating the farm from the school. A host of political shenanigans ensued, and it became clear that it would be years before any real change would happen. I would be stuck working with a difficult partner, and Joan wanted to return to Ithaca, where she had been offered a teaching position. After a year, we headed back to Ithaca. This time it was Joan who had the job. We assumed I would find my way.

We rented a house in Ithaca while we looked for a place to buy. It was obvious that I was going to have to change my profession, but I was not sure what it would be. I had no skills apart from farming. I did a bit of carpentry, but that ran dry. Joan's teaching job could hardly support us, and we were getting pretty low on resources, so I took a job with a cleaning company. I think it was during this time that I learned humility. Cleaning movie theaters and frat houses was the pits. Nevertheless, I learned how I could make good money by cleaning windows and having my own contracts, and after six months I started my own business. Sparkle Cleaning specialized in window cleaning. I also had a crew that cleaned a Hoyts movie complex and some office buildings. I was used to milking cows on Christmas Day, but cleaning sixteen movie theaters on Christmas Day did not have the heart that the cows had.

It took us a year to find a place to buy. It was out in the country with twenty good acres set up to grow vegetables. It even had a renovated house on it, just the right size for our family. I learned how to grow vegetables, and after another year phased out my cleaning business.

The trouble with growing vegetables in Upstate New York is that the winters are long, and during these months, there is not much income. I decided I would bake bread in the winters. I had read an article about building simple wood-fired brick ovens, so I went to a weekend workshop where they were building one. I also apprenticed at a bakery in Kansas, where they were using a wood-fired brick oven. I then went home and built my own oven with an eight-by-six-foot hearth in which I could bake sixty loaves at a time and do seven batches before the heat ran out. I baked mainly traditional European sourdough breads. The outside had a nice crust, while the inside was soft and chewy. My first few batches were a little on the flat side, but before long I got the hang of it and mine was soon considered the best bread in Ithaca. I had good markets in Ithaca, and I achieved my maximum, baking as many as nine hundred loaves a week. Fridays were my big day. I could sell a couple of hundred loaves at the farmers market, and the rest sold at Green Star Cooperative. On bake days, I would even deliver bread to the coop, right out of the oven. The whole store would fill with the smell of fresh baked bread and it would be gone in no time. My bread was successful and made a lot more money than the vegetables, so I gave up on them.

Eve and Dave were growing up and had become passionate about riding horses. We could not afford well-trained horses, so we bought two young horses that they worked with and trained after school. We bought an old horse trailer, and once a week we loaded up the horses for riding lessons. During the summer, we would spend Sundays at horse shows. Eve and Dave were good, but as they got older we couldn't afford the professionally-trained horses that they were competing against. It was a nice way for me to be with my kids. By tenth grade, we decided to stop, and they got into team sports at school.

It was nice to have a bit of extra money from the bakery that did not have to go into farm improvements. One summer, we closed the bakery for a month and vacationed in Europe.

We rented a car, visited friends, and camped. We started in Amsterdam and worked our way through Germany and Austria, down to Italy, and back up through France to Holland. I think Italy was our highlight. We camped outside Venice on one of the beach campsites and took the ferry into Venice, where we spent two days walking around the city and riding a Gondola. On the beach, we encountered our first topless bathers, which was especially interesting to Dave, who was fourteen.

Florence was wonderful, with all its museums and architecture, although the campground was not so great. However, the breakfast, fresh baguettes and coffee with the view over the city, made up for it. In Florence, the kids encountered their first toilet that was a hole on a concrete pad.

Another summer, we went out west for a month, rafted down the Colorado River, and camped and hiked at national parks.

During all this time, I was still holding onto my dream of getting back into dairy farming. I had the feeling that I had missed my life's calling and secretly wished that I would die of some illness. I was not depressed and told no one. I just felt that making eight or nine hundred loaves of bread a week, even if they were the best, was not what I was meant to be doing. Joan was happy, but when I told her how I felt, she accepted my needs. I do believe we create our own reality, and during those years I had to struggle with finding out who I was.

Christopher and Martina Mann, leaders in biodynamic circles, were looking for a biodynamic farmer to lease some of their land in East Troy, Wisconsin. I visited several times and, after much heart searching, decided to move again. We had been on our little farm outside of Ithaca for seven years and Joan felt very much at home there. She did not want to leave. This reminded me of a conversation I'd had at my first biodynamic conference in America many years earlier He was an older man, disappointed in his life and angry at his family for not letting him follow his dream when he was younger. He had owned a farm and, when they hit hard times, had to sell out. He had joined a dry cleaning business and then bought it, doing well. His dream had been to get back into farming, but his family would not allow it. Now he felt that his life had been wasted.

I think Joan knew that I would become that person unless she encouraged me to take on this new challenge. Eve was already at college, and Dave was going to graduate from high school in June, so this was a good time to move. I had to leave in March to prepare the Nokomis Farm for spring, and Joan followed after Dave graduated in late June.

CHAPTER NINE

MY HARDEST YEARS

Christopher and Martina Mann moved to the United States in the 1970s to help develop Anthroposophy and its sister movements of Waldorf education and biodynamic farming. They bought several farms in the East Troy area of southeast Wisconsin and funded Michael Fields Agricultural Institute, a research and teaching institution for organic and biodynamic farming. They reorganized their farms because there had been no cows on the land during the past few years, and they felt that the fertility of the land was going down. They asked me to lease their dairy, and we came to an agreement that I would lease the land, and they would put up a set of new buildings.

I was then able to build my dream facility. We built a milking parlor, a double-sixteen swing parlor that was very efficient and could milk up to a hundred cows an hour. There was an adjoining office and space for a classroom. The cows were housed in a big loafing shed, where they could run loose on a pack of manure and clean straw. This method produces lots of compost and is very easy on the cows. The facilities were excellent, and I was very appreciative. I found a herd of certified-organic cows nearby, and by April we were milking cows. I ran the farm business as my own, so I owned the cows and machinery and had to take out loans for about one hundred and forty thousand dollars, which should have been quite doable. I was excited to be running my own farm again and barely had time to miss Joan, who would be able to come out at the beginning of July.

Transitions were always difficult for Joan, and although she was not happy, I did not think too much of it. She was going to start teaching first grade at the nearby Waldorf school, and I reckoned she would make good new friends, as she always did. However, once she got to East Troy, she would wake up in the mornings crying and feeling that this was not her place at all. She was always brave, so she carried on making the house and garden into her new home and preparing for the school year.

The day came for Dave to start college at George Washington University in Washington, D.C. I would fly to D.C. with Dave to get him oriented, so we had a two-hour drive down to O'Hare airport. Meanwhile, Joan had to attend her first faculty meeting, so we all left about the same time, with Eve still in the farmhouse.

When I got to O'Hare, I was being paged, which surprised me, since I had our tickets and everything else in order. I left Dave at the check-in line and found the police office, where they informed me that Joan had just died in a car accident, though they had no details. I had to go back to the line and tell David that we were leaving the airport because his mother had just died.

It was a terrible, long drive, just Dave and me, not knowing anything about what had happened. Eve was waiting in the house alone, since nobody else knew. After returning home,

I had to call for help and tell people. The community rallied around and helped the best they could. Friends and family came from great distances. My sister flew in from New Zealand, her daughter came from Japan, and my brother Johannes and his wife came from Ireland. Joan's brother arrived from Long Island, and her parents from Florida. It was especially hard to see her father's grief; his only daughter was gone. Old friends from Kimberton and Ithaca arrived, too, and we were joined by many new friends from East Troy.

The Christian Community priest, who now lived close by, was especially helpful. He had accompanied Joan and me in our life, had married us and christened Eve and Dave, and now took care of all the funeral arrangements. I remember he took me to the funeral home that afternoon so we could arrange for Joan to be taken home rather than stay in the funeral home. I was not allowed to see her because she had been so badly hurt. Joan had run a stop sign, not even slowing down, and had been hit side-on. She was hit so hard that her watch stopped and she was knocked out of her shoes.

During the three days before the funeral, we were all sustained by the outpouring of love of family and friends, and it became a vibrant celebration of her life. During the day, people would visit and we often had music. The evenings were more intimate, when family or close friends would share stories of her life.

After the funeral people left, and there was a depth of loneliness that was at times unbearable to me. Eve and Dave decided that it would be best to continue with college. Dave and I went again to O'Hara and flew to D.C. to get him settled into college. A week later, Eve left for her four-month study abroad in Africa. Luckily, my sister stayed on for a month and helped me through some of the worst part. When my father had died a few years earlier, I had been at his side and experienced his "being," or soul, leaving and expanding into the universe. This had been very special. Unfortunately, however, with Joan there was nothing similar to give me solace. She was gone and I could not feel her presence. After a while, the emptiness was not so constant, but the grief would hit me hard at unexpected times. It would usually come at times of beauty. Seeing a flock of birds changing directions in mid-flight or the wind blowing through a field of wheat would leave me feeling desolate and in tears.

About two months after Joan died, a friend passed away and I went to the funeral. She had died of cancer and had a long time to prepare. She and her friends had written the funeral service and she had felt that, after she died, she would be experienced in the wind, in the sunset, and in other aspects of life. I left the service furious and shaken to the bone because, for me, there was nowhere to go to experience Joan.

Working during the day was okay, because I had to keep the farm going and there were things to keep my mind and body busy. The nights were much harder. When I closed the door to the milk house in the evening, I dreaded going into the house. I would sit on the steps and be with the cats and dogs for as long as possible before going in to make supper. Often at night, I would lie on the floor, light a candle and listen to Handel's *Messiah*. This music let me experience death and then resurrection, and it helped my healing process more than anything else did. Saturday nights were especially difficult, sitting by myself watching TV.

I am not the type of person who likes to sit around talking with a group of people. I prefer one-on-one conversation, so I would often feel very lonely, even within a group setting. I knew that Joan would not be happy watching me feel sorry for myself, so I decided to take the plunge and invite a woman out to dinner. I knew one of the Waldorf schoolteachers who was single, so I checked with a mutual friend about whether she would be open to going out to dinner. Happily she was.

I had not dated for twenty-five years, but I reckoned we could talk about teaching, and Celia had even lived in New Zealand. We had a good conversation and became friends, and later she moved in. Following one of the first times that she had supper at my house and had left, my whole body went into a panic. The thought of her leaving left me in tears, and I realized I had an irrational fear that I would never see her again. Celia helped me get through those lonely months, although I know local people were upset that I found a new partner so soon. We were together for two years, helping each other through some hard times. She was twenty years younger than I was, and I realized I didn't need another daughter. Likewise, she realized she didn't need a father figure. When the time was right, she moved to Oregon and soon found the man she would marry.

At the time of Joan's death, a friend had given me a moth chrysalis and mentioned that it might hatch in six months. I left it in my bedroom and did not think about it. One night I walked up the stairs to my bedroom, where I found a beautiful huge moth sitting in the middle of my pillow. For me it was a gift from Joan. It meant that, although I could not experience her at that time, in the future we would again have a relationship that would transform into something new and beautiful.

Eight years later, I was at a workshop led by Kimberly Herkert, cofounder of Way of the Heart. One of the sessions was on forgiveness. Joan came into my mind, and I became upset that she had left me holding the bag. I had felt that we'd had a lifelong agreement to support each other. Of course, this feeling was irrational, and I didn't even know that I carried it, but I could not forgive her at that moment.

At the end of the session, we went round the group, each person briefly telling of their experience. I was nearly in tears, and it was hard to speak and talk about my feelings. Kimberly looked at me and told me that I had just forgiven Joan. That evening, I went to the beach and did indeed feel like a weight had been lifted off me. Since then I have again felt closer and more at ease with Joan.

CHAPTER TEN

SUSAN INTRODUCES ME TO MY KINDRED SPIRITS

When Joan died, I had just turned fifty and felt like I was being forced to start a second life. I had lost my wife, the kids were grown up and off to college, and I was starting a new business in a new town. After Celia left, I was able to face life on my own, and day-to-day life seemed more manageable. However, I do not think I was meant to lead a bachelor's life.

Susan and I met one day in the farm yard when she was there for the board meeting of a children's program on my farm. One of the children had accidently locked her keys in her car, and I noticed that her car window was opened a crack, allowing me to use a coat hanger to open the door. For my help she promised me an apple pie, and when she arrived with her gift, we found we had much in common. Not in our work, because I work with the land and she works with influential people in social investing and social ventures. However, we were both trying to transform our world—she in social investing and I in biodynamic agriculture. Nevertheless, she also felt a strong connection to the land.

Susan had been married to a Nigerian who was both chief and a healer of his village. He had received a Ph.D. degree from Harvard, had been the number-two man in Nigeria with the first democratic government, and had lost his wife and children in a tragic plane crash. At that point, he had taken up duties as the key village elder of his tribe and, after that, married Susan, his college sweetheart. When Susan joined him, she shared the subsistence lifestyle in his farm compound and learned that it was possible to live off the land in simple surroundings. After six years, however, the cultural differences had proved too hard to bridge, and the marriage ended. Susan returned to the US heart-broken. Soon after, we met, and our similar experiences of the grief of death and separation gave us a deep kinship.

When Susan arrived with the apple pie, it was really good, so I invited her to the Milwaukee Symphony. After the concert, while waiting for the parking garage to empty, we went for some coffee. She had never studied Anthroposophy, so I offered to read *Theosophy*, one of Steiner's foundational books, aloud, after which we could have a discussion. She in return offered me dinner, so it was a good deal for both of us. After work, I would arrive at her house for dinner and our discussions, and they grew more and more interesting. Looking back, it was strange that I offered this, since I am not the type to have an intellectual discussion about a book, but I had to know if we would be compatible in our paths.

During the next two years, we got to know each other better and better. It was a very enjoyable courtship, because we could afford to go on romantic vacations. We spent three weeks

in New Zealand, mainly
touring the South Island in
a camper van. I had remembered many of the underde-
veloped camping sites from
my youth, and we would
search them out and camp
in our van. In the morning,

Our wedding

we would wake with the Sun rising through the mist of a nearby
river and the Southern Alps as a backdrop. Not a person in
sight; just a cup of coffee to warm us up. Another time, we
visited England and stayed at bed-and-breakfasts. We had a
general idea of the places we wanted to visit. Oxford for a day,
London for a few days, Stonehenge, my old boarding school in
Sussex, and places where we stopped on the spur of the moment.
We were not on a tight schedule, so we could relax and get to
know each other.

In the meantime, I had a farm to run. This was a real challenge, because I could not get the cows to make enough milk.
They had many health problems, such as bad feet, low concep-
tion rates, and high somatic cell counts. I could not figure out
what was wrong until one day a friend suggested I had a stray
voltage problem. Cows are unbelievably sensitive to voltage dif-
ferentials in their surroundings. It stresses them out and causes
their immune system to kick in on a permanent basis, leaving them little resistance to fight disease. It took five years to
solve the problem! A veterinarian who was also a dowser came
and tested the farm. He found electrical earth currents going

through my milking parlor. It was amazing to see his rods turn when he crossed between a low-lying pond that was picking up stray voltage, the electrical control panel for my barn, a drilled well just outside the barn, and then to the transformer. We solved the problem by constructing a medicine wheel from fieldstones off to the side of the milking parlor. Through dowsing, we were able to place the medicine wheel in the right place so that the earth currents could go in a different direction and not affect the cows.

I know stray voltage does not sound devastating to non-farmers, but it cut drastically into my bottom line. Every year I sold about $250,000 worth of milk, but stray voltage was causing my cows to drop ten pounds of milk per day, which added up to a loss of $50,000 per year. Because of the stress the cows were experiencing, I had a high culling rate (cows no longer giving milk) of more than thirty-five percent, so that each year I had to buy heifers that cost $1,500 each. I was cash-strapped and had to refinance several times. On top of that, I experienced a string of three drought years, which made it necessary to buy a lot of feed. One year, I was trucking in certified organic hay from Montana and Kansas. The hay cost $800 per semi-truck load, and the trucking came to $600, while a load would last only ten days.

At times life became nasty. I would get phone calls from my suppliers, saying that they would not deliver without a check. I would have to put off paying dealers, who would then charge eighteen percent interest. I would take short cuts that got me through a month or two but hurt deeply long term. It also hurt my reputation as a farmer, because I would have to postpone repairs and maintenance. For example, a farmer is obliged by law to keep his thistles mowed, but I could not afford the right mower, forcing me to beg the neighbor for the use of his mower.

Sometimes I would catch up on my payments, but it was very stressful, and I worked long hours. I thought I could be like a duck and let the rain slide off my back without hurting inside. But I did hurt, and eventually it caught up with me, and my health deteriorated. My muscles became like cables, though nobody could diagnose the condition. I had to cut my hours way back or I would feel my back muscles tensing and preparing to pull my back out. I sensed that I could not even let myself get angry, because the adrenaline it pumped into my body would leave me aching all over. That is a strange experience...trying to be happy when a cow shits on you in the milking parlor.

Soon after I solved the stray voltage problem, things started to improve. My calves did not die, so I was able to raise all my own replacements, and my culling rate came way down, although it took a few years for the older cows to respond. I had some cash to spare, so I was able to replace some of my old machinery and catch up on maintenance. During that difficult time, I continued to believe in myself. This is a spiritual lesson in itself—to know that you are on the right path despite the adversity. I also felt a lot of criticism coming from the community and that I was put in the box of a failed farmer. I went through an initiation by fire.

I felt especially proud about my cows. I did several things out of the ordinary, like not using artificial insemination to get the cows pregnant. I had crossed my initial herd with Normandy bulls, and in eight years had created an all-round cow that did well under grazing management. I had decided that I would leave the calves on their mothers for four weeks rather than take them away at birth as others do. I could see that the cows craved to keep their calves, for when I started the practice, the other cows would gather around the newborn and not leave the mother and calf in peace. Some cows had sneaky ways to steal a calf from its real mother, which was bad, as the calf would not get the colostrum milk it needed. After a few months of this practice, when a calf was born there would only be mild curiosity on the part of the other cows. At first, it was difficult, as the mothers had lost much of their mothering instincts, and it was common to lose calves out in the field. After two generations, the bonding between mother and calf returned. I felt it was important that the mothering instinct of the cows be respected and that they be allowed to fulfill this basic instinct that they craved so strongly. To see a cow and her calf together is truly moving. Even keeping the bull with the herd made a difference, as it made the whole herd less nervous and more contented.

When I first designed the farm, I had made some false assumptions. As a model to design my farm, I had used a nearby farmer who was a grazier and used the biodynamic preparations key to biodynamic farming. I assumed he was trying to create the same kind of self-contained farm organism that was so important to me. In his scenario, he needed two acres per cow but he bought in all his concentrate feed. I, on the other hand, wanted to grow all my own feed, not just the pasture and hay, but also the corn and beans, so I actually needed four acres per cow. I was locked into a facility built for 120 cows, but only had 240 acres. Over the years, I was able to find another 260 acres to rent, but it was a struggle to farm so many acres. Looking back, I can see that it would have been better to design a set of buildings and put together a budget more appropriate for 220 acres.

I continued to hold my dream of helping people experience ways a farmer can steward the Earth in a caring and non-exploitive way, while producing good food. Over the years, Susan had created a network of close colleagues who were interested in social issues, which often included responsible land stewardship. She invited fifty people to join a network that we called Kindred Spirits (see kindredspiritsnetwork.com). For this, we charged one thousand dollars per person, which helped our bottom line. We invited eight people at a time to come and stay at our farm for a long weekend. There were two main themes worked into the stay. Susan has a gift of matching people so that it would be comfortable for each person to talk about their life and how they were fulfilling their destiny paths. From these discussions, they would get encouragement and support in their life decisions. My part was to take everybody for a walk through the farm and teach about how non-farmers can steward the Earth. We would end up on our sacred hill, where I would talk about biodynamic farming and the spirituality of the Earth. Many people have lost contact with farming and do not have a chance to experience farm life. Yet they know that their very sustenance is dependent on the Earth, so they appreciated this chance to see where their food comes from. In December, we would have a weekend for all fifty members, but they would stay in a nearby hotel. As Susan had carefully chosen the group from her life's work, the conversations were substantive and revolved around the idea that humanity has its ladder up the wrong wall. People think that the environment is a subset of the economy but actually, the economy is a subset of the environment, because it is the Earth that supports us all. In farming in particular, we can make decisions most consciously that affect both our health and the Earth. In biodynamic farming, we have the added dimension that we work with spirit that stands behind nature. Through Kindred Spirits, Susan and I found a way to be and work together and share our lives with our friends. It was also a Community Supported Agriculture (CSA) project but instead of giving away vegetables in return for financial support we gave away the opportunity to learn how to steward the Earth and to have destiny path conversations with a like-minded group of people. I think the fifty people did feel a deeper connection to the Earth through our farm. We created a network that covered many walks of life and covered the states. Two members in particular captured the value of Kindred Spirits. On their website, master chef and authors Karen Page and Andrew Dornenberg wrote, "We came to one of the smallest villages we had ever visited to hear some of the largest ideas we had ever heard."

After eight years, I decide to give up my lease. This was hard but I was ready for a change. I have always been willing to step into the void. I wanted to find a way to talk about farming and my love of the Earth. Kindred Spirits had allowed me to experience this possibility, but in the states, I could not see this opportunity opening up. I needed time to recharge my batteries and to deepen my connection to nature. Serendipity soon gave us a path through the void.

CHAPTER ELEVEN

A VISION QUEST IN ECUADOR

Susan and I had already made a connection to Ecuador and built a small house. Now we had an opportunity to live there. I was lucky in the timing; the market for organic milk was still growing. I sold my cows for a very good price and was able to recoup my losses. I had no plans to get back into farm ownership when we moved to Ecuador. However, by true serendipity, I was offered a twelve-hundred-acre farm, which I was able to buy with the money from selling the cows. We named it Serendipity Farm. The land had some of the purest air, water, and soil on Earth and represented true wealth. We do not live on the farm, but once a week I go out and make sure everything is okay. Andreas, a young farmer, runs it for me, and we have a wonderful relationship. It is great to have a farm and not have to worry about the cows getting out.

Farming is very low-tech here, so I have to rethink much I have learned. Andreas grew up on the farm with ten brothers and sisters, and it was his dream to run it one day. However, he never had the capital to buy it from his mother and support his young family. I pay him three hundred dollars a month, which is on the high side of average. This made it possible to buy thirty

The author with Andreas at Serendipity Farm in Ecuador

beef animals and pay for capital improvements such as fencing and extending the watering system. What makes it financially viable is that I can sell a full-grown bull for six hundred dollars, a bit less than I could get in the US, but my expenses are so much less: $300 for monthly wages instead of three thousand plus other extra costs of being in the States. Moreover, everything is done in cash. My property taxes are $16.68 per year. My house is a bit more, $48 per year. On the down side, bureaucracy is a nightmare. There is no mail delivery; we still have no mailbox in Loja, the main city an hour away, so it took three trips to Loja to be able to pay my taxes. The best way to do business is to smile widely and apologize for my poor Spanish, while I learn it as fast as I can. *"Desculpe me, yo hablo un poco español"* (Excuse me, I speak only a little Spanish!) goes a long way. People are very kind and patient and I do my best to be kind and patient in return

The author at a neighbor's farm in Ecuador

We have other projects that help us integrate into the community. Before going to Ecuador, a philanthropist friend of Susan gave her $20,000 per year for three years to give away for the highest good (flowfunding.org). There are a few healthy conditions attached. The money cannot be used to pay oneself or one's relatives or one's expenses and it cannot be used for personal projects. Susan was so touched by the generosity that she decided to create one of her KINS Innovation networks pro bono. She let the members give out the money, because local people would know how it could do the most good. After six months of ferreting out collaborative people in different sectors in Ecuador, she started a little network called *Ayni,* the Ketchua word for reciprocity. Eight of us (four Ecuadorians and four foreigners) get together about once a month to discuss how the money could best be used to help the community.

For instance, in the valley below our house is a very poor indigenous community of the Saraguro people. The young women had asked that we pay for a dance instructor so they could learn their traditional dances. An Ayni member decided to support this project pro bono, and I help him. Every Saturday I would pick up the dance instructor (herself a student at the university in Loja) and watch the women practice for two hours. If neither my colleague nor I are there, the young men would disrupt the lessons. At first, the girls were very shy and self-conscious, but they gradually got over it. Usually I take a book along to read, and occasionally I'd nod off. That brings a smirk to their faces, and it brings us to the same level. The women are very suppressed by their men, and the dancing has given them a new belief in themselves. They came in second at the Vilcabamba carnival parade.

Now that my Spanish is improving and I can have simple conversations, I am starting to make friends with them. All together, Ayni has more than twenty-six projects that are carried out pro bono by the members, with all out-of-pocket expenses covered. They range from a "Pay It Forward" program to the free healing offered by our leading shaman, to covering the cost of rebuilding materials when people have lost their houses to landslides. In all, there are more than two dozen Ayni projects in various stages of development and the total out-of-pocket costs after three years has been $60,000.

Vilcabamba has many interesting people and we enjoy their company, but mostly I stay at home meditating, reading, writing, and working with our gardener and the cows I run on the Finca VIVA land surrounding us. Susan is the people person, so

A cow and calf at Serendipity Farm

she always has many people to meet. She is a driving force in getting the waste disposal problem fixed through Ayni. All of Vilcabamba's waste is going to a site just above our house where it was being burned. If the wind was going the wrong way, we would be sitting in a cloud of toxic fumes, and then it would drift down to the village. This was rather upsetting, as we had traveled thousands of miles to be in a clean environment, and all around us there are thousands of miles of clean Andean mountain air. Now, with the help of Ayni funds, the garbage is no longer burned and a report showed that eighty percent of the garbage is organic. Soon a recycling program will be started which will create lots of organic compost which will be given away to local farmers to encourage them to try organics. The non-organic garbage will be recycled or taken to a proper landfill in Loja. We must also mitigate the existing open-air dump. While estimates of the mitigation were a million dollars at first, by locals and foreigners collaborating, we are hoping we can do it for $6,500.

Susan worked intensively on the Ayni projects but also found time to write her book, *The Trojan Horse of Love,* and finished it literally the same day I finished this book. The synchronicity felt quite amazing to us. Her book describes the twenty KINS Innovation Networks she has created over the last thirty years in social investing, solar, corporate social responsibility, organics, micro-enterprise, and women's empowerment. These networks bring servant leaders in each field together to collaborate through their higher consciousness. As a result, they have been incredibly effective. She also describes her joy in trusting her intuition to take risks around her values to find her destiny path. Rather than selling her book, she gives it from her heart to others and encourages them to give it from their hearts, too. (Visit thetrojanhorseoflove.com, the web site for the book.)

About a month ago, my back went out again in a more serious way. It was like old times in East Troy, where I had to lie on my back for three days, with every movement being excruciating. This surprised me, as I know that my back goes out only when something is bothering me, and life had been quite serene here. In fact, my state got so bad that I developed a fever and felt nauseous. Carlitos, the shaman healer from Vilcabamba, visited and told me that I was on "a vision quest at home." I knew this was true. For two years, I had been trying to live in the moment and to still my mind. As I mentioned, when I sit on our patio early in the morning and watch and listen to nature, I start to feel light and part of the formless all. I lose my mind identity, which is a wonderful experience. It seems my body was not agreeing with this assessment. Over the years, we put our

The cows on my mountaintop

stresses into our bodies, and for me I seem to put them in my lower back. I went through a crisis with this back pain and was determined to get to the bottom of it. During my second night of being unable to move, I was able to break through. It really is hard not to worry about the future based on experiences. It is hard to believe that my value does not lie in what I have accomplished or will accomplish in the future.

Help did come. I was able to detach myself from my life and to go to the mountaintop spiritually. I looked at myself growing up and pursuing my dreams. I saw how I lived a life full of joy and sorrow, but a wonderful life, living with what I had inherited, both good and bad. I had wrestled with the earth to make a living and in the process, I was molded and taught, despite my hardheadedness. The Earth embraces me, loves me, and allows me to fulfill my destiny.

As I write this, my back is still slowly recovering, and Susan and I had to celebrate our fifth wedding anniversary at home instead of hiking the Podocarpus National Park nearby as we had planned. However, my bad back served its purpose, because I am planted more firmly now in the present, trusting the spiritual world more deeply than I ever have before. I am fulfilling my intention to spread biodynamic agriculture by writing this book ... and I hope readers enjoy the coming chapters on biodynamics for non-farmers.

My Spanish is now getting good enough that I will soon be able to give talks on organics to local farmers. I will find ways to fulfill my intention of teaching non-farmers how to steward the Earth, offering talks and walks here on Finca VIVA and in other locations. It is most important that I have moved into a deeper level of self-trust, of staying present, and of receiving help from the spiritual world. That is enough.

We are so lucky to be on this beautiful Earth. I feel lucky that I was led to farming. I am excited despite all the challenges, the flies and manure, the kicking cows, the broken-down machinery, the draughts and floods, and all the work with little financial reward. At night when I look at the uncountable stars glowing overhead, I experience the infinity of peace and love and wisdom of the spiritual world. Out of spirit, all this has risen. In all humility, I have been allowed to take part in the alchemy of creation and destruction through farming. All around me, I see and experience this beautiful world and know that I am blessed ... that all creation is blessed. This is my love, a farmer's love.

CHAPTER TWELVE

BIODYNAMIC FARMING
MY LIFE'S WORK

As part of this book, I will describe what biodynamic farming means to me. Every biodynamic farmer would probably emphasize different aspects, but this is what is important to me. I have tried to describe the farm ecosystem, as well as how I attempted to create this on my last farm. It is especially hard to write without using anthroposophic terms about the various levels of the spiritual world and how they manifest in nature, because one must assume the reader is not at all familiar with Rudolf Steiner's teachings on the subject. Biodynamic farming can be a lifelong study, and I still enjoy visiting farms and picking up new ideas. I am still learning and refining these concepts, and therefore I decided to add these chapters at the end instead of interspersing the ideas throughout the book.

THE GREAT ARTIST OF THE LANDSCAPE

The spirituality of the Earth has always been important to me. As I mentioned, when I was nine, walking home through the bush, I experienced losing my oneness with nature. By farming my whole life, I could at least be out in nature and enjoy myself.

Trying to make a living from nature has been hard, but I have always had my spiritual perspective to keep me going. When most people think of nature, they think about a secluded place or time spent in a national park. For me, nature is all around me when I farm. I like to think of a farmer as the great artist of the landscape. Every decision we make changes the look of the land. As human beings, we impinge most on nature where we grow our food, and overall we have done a terrible job. Just think of the corn and bean farms of the Midwest, where people are literally unwelcome. Not only are they dangerous places to visit because of known chemical hazards, but there is also no place for human beings there. The farmer could show you his or her farm, but it would be in a pickup truck in a cloud of dust along endless rows of corn or beans. Even worse are the chicken houses, the beef lots, and the huge dairy farms.

A biodynamic farm, by contrast, is diversified; it is interesting and beautiful. It is a place where people like to visit and feel welcome. Not only does it grow food that nourishes, but also allows people to feel connected and safe.

I was lucky that I have been able to farm the land and grow good food. My late wife Joan, with the help of our two children, had the social ability to welcome people into our home and farm. Later, Susan and I married, and she had that same social capacity. In particular, she guides people to find destiny paths... and what better place to search than on visits to a biodynamic farm? Thus, on our farm in Wisconsin, we invited her friends to visit for the weekend as part of our "Kindred Spirits" network. I would take them walking through the pastures, and in one I would invite them to sit in a circle as the curious cows would gradually gather around us. There I would talk about the spirituality of the Earth and biodynamic farming.

As I wrote earlier, the turning point in my life took place when I was going through some very hard times in my marriage and the most incredible being of light and love suddenly visited me. I experienced the greatest wonder and appreciation of everything that I had done in my life and felt understood, accepted, honored, and loved. Many years later, I had a similar experience of love, though not quite so intense.

As background, let me explain that, for me, cows are part of the Earth. The being of the cow, in her loyalty to the land and the cosmos, belongs to the landscape. One late August afternoon, I was getting the cows ready for milking and they were being stubborn. It was hot and muggy and I was irritated because I had more hay to make. As I walked past one of the cows, I happened to look into her eyes, and we began a deep conversation. For my part, I said "I am sorry, please forgive my irritability, but I've got problems." In reply, she communicated back to me incredible forgiveness and love. I experienced the Earth welling up through this cow. The Earth, as a being of light and love, came shining through the eyes of this cow. I was startled, yet deeply moved that in my frustration I was allowed to experience this union with the Earth.

Thinking about this later, I realized that it was the same love I had experienced when I was thirty-three and the being I think of as Christ visited me. I understood then that the being of Gaia is now permeated with light and love, and that this light and love are being extended to all humanity now in an unlimited fashion.

Out of this stems my ever-growing love of nature and of the being of the Earth imbued with love. My path has not been a scholarly one but more a life of doing. My main inspiration has always come from Anthroposophy, but often there has been little energy left in the evenings to study. One of the nice things about being a dairy farmer is that you are forgiven if you fall asleep at meetings. Now that I am not farming and can be more awake, I will share how my experiences allowed me to see the spirituality of the Earth and how biodynamic farming led me to my worldview.

My experience of the Earth being imbued with light and love is further confirmed by meditation. When I look deep into the Earth in my imagination, I move through matter and experience the Earth as hollow, surrounded at the periphery by light-filled crystal, dissolving into darkness. The hollow Earth itself emanates light and love. The first time this happened I was surprised, as I expected density, weight, matter, and gravity.

In my reading from Anthroposophy, mainly Sergei Prokofieff and Jesaiah Ben Aharon, these imaginations are confirmed. To me, they are attuned most to the changing Earth. In addition, in a course on geomancy with Marko Pogacnik, I started actually to experience the spiritual landscape that underlies the physical.

The being of light and love that I experience is personal and present, but also historical and cosmic. For me, this being is the Christ Spirit, the beloved one, who has accompanied the Earth and humankind from the beginning of time. This is the god that ruled from the Sun realms, so all peoples have venerated this being in one form or another. The Egyptians called him the mighty Sun God Ra; the Greeks called him Apollo. Slowly, as our consciousness descended from the dreamy clairvoyant awareness of spiritual reality to our more sharply defined sensory perception of earthly reality, the Christ Spirit drew closer to the Earth. He incarnated into the person of Jesus and then united his being with humanity and with the Earth. Through that act, he made his new home on the Earth for all time to come. This was a gift from the spiritual world, as we had lost our connection to spirit.

In the distant past, our way of being was spirit-imbued. We beheld and experienced spirit in matter. Today, when we think about nature, we experience an abyss. We cannot cross the bridge between matter and spirit. When we see a tree, we see only the physical tree. We do not see the spiritual tree, imbued with life force, or the spiritual beings that surround the tree. People with spiritual vision, however, do see them. Now we are starting a new era, when our spiritual organs of perception are reawakening so that more and more people can again see the spirit in matter. We are starting to see the etheric world with new spiritual sense organs. In the future, we will be able to understand this realm and then be able to co-create with it. Even now there are forerunners. The Findhorn community in Scotland has been creating an oasis where none would seem possible. By taking direction from the nature spirits, they have miraculously created a lush garden out of sand dunes.

As time goes on, over the next several thousand years, people will experience and live into this realm Anthroposophy calls "the etheric." Already some people live without food by tapping into spiritual energies. The physical will gradually be less able to support us. My favorite grace expresses this so well:

> The bread is not our food.
> What feeds us in the bread
> Is God's eternal word,
> Is spirit, and is life.

It is the spiritual forces in the food that sustain and nourish us. This is why biodynamics is so important to me. By looking into the spiritual world, Rudolf Steiner has given us a way to grow food with the spiritual forces that are necessary to enrich humanity.

CHAPTER THIRTEEN

STANDING BETWEEN HEAVEN AND EARTH

The biodynamic farm stands between Heaven and Earth. When looking after the land, a farmer tries to work and balance those influences, the cosmic and earthly. Rudolf Steiner showed us how to work with the celestial forces through the use of the biodynamic preparations and by being aware of the position of the planets in relation to the Sun and Moon and the constellations of the zodiac. When we think of the terrestrial forces, we have in mind the physical properties of the farm, mainly the soil.

The smell and feel of soil tells much. Is it rich and earthy with little bits of organic matter still visible? Does it feel full of life? Does it crumble between the fingers? Is the soil deep? Are there lots of earthworms? Can the soil hold moisture? These are all good signs of a fertile soil with lots of organic matter. What is the bedrock; is it clay or sand? What is the vegetation? Are there trees and hedgerows? Is the terrain steep or flat? These are all terrestrial forces that a farmer works with on a daily basis. Although we cannot do much to change some of these things, such as the original parent material of the soil or the slope of the land, over the years we can modify and improve the fertility.

In the past, peasant farmers felt the holiness of the Earth, but today our large-scale farming and economic pressures make it hard to maintain a feeling of reverence. It is interesting that in Rudolf Steiner's *Agriculture Course* he spent considerable time thanking the hosts of the conference. This sets the mood for farming. Reverence used to be a basic mood of soul when working with nature. Growth and decay, birth and death, the wonder of a newly planted field greening up all bring me a feeling of wonder and gratitude. One naturally starts to see the world in a flow of time and movement. This leads the farmer through a path of initiation that can make it possible to hear intuitively what the farm needs.

However, such a natural path of initiation is not necessarily available to a modern farmer. Sitting in an air-conditioned tractor cab and listening to the radio to keep awake for hours on end... this is not conducive to a spiritual path. One now needs to exert a conscious inner spiritual effort to gain an intuitive connection to the farm. By cultivating the inner soul life and caring for the land and animals, a farmer can develop an intimate connection to the farm. The farmer is like a mother who is connected closely to her small child and knows when the child is in danger. Even in my thirties, when I was most involved in building up my farm and most active with physical work, I would try to put aside thirty minutes after breakfast for meditation.

In addition to the terrestrial forces, cosmic or celestial forces also affect the growth of plants and the health of the animals. The Sun is the driving force of our planet and allows the great variety of plants and animals ranging from the poles to the equator. The Moon affects mainly the watery aspect of our world. This can be seen in the rise and fall of the tides and in the growth habits of plants. Before the full moon, seeds germinate and grow faster than at the new moon. It is easier to make and dry hay at the new moon, when there is not so much moisture in the stems and leaves. We know of these influences but are mostly less aware of the more subtle effects.

The energy that reaches the Earth changes as the Sun stands in front of each of the twelve constellations. The growth and nutritive value of the plants is dependent on this dance of the heavenly bodies. Rudolf Steiner gave us the gift of being able to work with the cosmic forces through the biodynamic preparations. With the use of the preparations, we can enliven the Earth so that our food can have the nutritional forces we need. When Steiner was asked why people were not more successful in their intentions for spiritual transformation, he answered, "Nutrition as it is today does not supply the strength necessary for manifesting the spirit in physical life. A bridge can no longer be built from thinking to will and action. Food plants no longer contain the forces people need for this."[1]

One of the main concepts of biodynamic farming is that we try to create a self-sustaining farm organism. This idea is

---

1. Steiner, *Agricuture Course*, p. 7.

wonderful to me, because it lifts the farm out of the economic realm into a cultural, artistic realm. It creates boundaries; we have a picture frame on which to balance all the terrestrial and celestial forces at work within this planet. From above, we have the influences of the Sun, stars and planets. On Earth, we create a vessel composed of soil, plants, animals, and the farmer, who creates the vision and orchestrates all the parts. All living entities have a skin that embraces the organism, whether they are unicellular or as complicated as a plant or animal. A farm is part of the Earth, and it is part of a farmer's task to create a living system, a landscape with the right balance of plants and animals so that a vibrant, self-sustaining whole is created. A modern conventional farm has no boundaries; it is bound only by economic considerations. Its very concept is unhealthy, since there are no limits to stop growth and therefore growth can become cancerous.

On Nokomis Farm, my last farm in Wisconsin, I tried to create such a diversified landscape. It was a beautiful farm, with areas of woods reminiscent of the original oak savannas of the plains. There were marsh areas with small streams that the wildlife loved. There were hedgerows between pastures, and Susan always said that our farm was a bird sanctuary. Within one area of woods, on the top of a fine hill, we cleared the grasses for our gatherings. Kindred spirits would gather there when I talked about the spirituality of the Earth, so we called it the Sacred Hill. Many ceremonies were held there for our friends and for the community, and Susan will never forget the circle of women who helped her celebrate there on the morning

Fields in the fall at Nokomis Farm, East Troy, Wisconsin

of our marriage. On Nokomis Farm, we felt the animals, plants, people, and spiritual world were one.

I had some areas of good fertile soil, but most of the farm was poor, rolling land, good for pasture and hay and dairy cows, but not suitable for intensive crop farming. I farmed about five hundred acres and needed to have about four acres of pasture, hay, and crops for every cow. Therefore, everything was designed for one hundred and twenty cows. Through experience, a farmer learns how much feed a cow needs per year to attain a reasonable level of milk production. How many tons of hay or silage are needed? How many pounds of concentrates such as beans and corn? This has to be converted to acres to be planted for each crop so that there is enough to feed the cows through the long winters. Then this is balanced against the tons of cow manure produced. Will there be enough compost for the

Normandy heifers on Nokomis Farm, East Troy, Wisconsin

necessary acres of corn? This can never be an exact science, since the situation changes yearly according to the weather. It depends very much on when the rain falls. If it is a wet, warm spring, it could be that the corn gets planted late and cannot be cultivated to kill the weeds. Yet all this rain is excellent for pasture and hay. If it stops raining for a couple of weeks at hay-making time, then a bumper crop can be harvested. One summer the situation was terrible, because it never rained and I was buying hay by October. That year I was certainly far from my dream of the farm being self-sufficient.

The rotation that worked best for me was forty acres of corn, forty of soy beans, forty of small grains (mainly oats but it could be rye or wheat) under-sown with hay that would last three or four years before the cycle started again. This utilized about 280 acres, and I had another 200 acres in permanent pasture. In addition, my farm provided manure for a neighboring CSA that grew enough vegetables for 120 families and other markets. I also sold wheat to a kosher bakery, and sometimes I was able to sell the beans for human consumption.

The farm organism does not stop at the farm gate. It extends into the community via the customers. I sold my milk to Organic Valley, one of the largest certified organic distributors in the country, which caused me to lose my direct connection to the consumer. However, it is still interesting to see how much food we could produce on my farm with just three workers.

Milk

Being a dairy farmer, I produced mainly milk. Each cow produced forty pounds of milk per day, or 1,500 gallons per year. An average family consumes about three gallons of milk and milk products (cheese, butter, yogurt, and ice cream) per week, or 150 gallons per year. This means one cow supported a hundred families, and since I milked 120 cows, we supported about 120 families.

Meat

A byproduct of milking cows is that you get bull calves, which are usually sold to another farmer, and culled cows that are sold for butchering. We sold about 20,000 pounds of meat per year and, according to statistics, beef consumption per person is about sixty-five pounds per year. If a family of four eats two hundred pounds per year, then we supported about a hundred families.

Wheat

I grew about twenty acres of wheat per year with a yield of forty bushels per acre. This averages 48,000 pounds, or enough for 24,000 two-pound loaves per year. If a family eats two loaves per week, we grew enough wheat for about 230 families.

Vegetables

Although we did not grow vegetables beyond Susan's garden, the cows supplied much of the fertility through their manure, and our vegetables were grown on the farm property. The twenty acres of vegetables that were sold through a CSA, farmers markets, and wholesale channels were enough to feed 300 families.

From this description we can see that a farm with the Sun and rain and the work of the farmers can produce a truly remarkable amount of food. For Nokomis Farm alone, this totaled:

Milk – 120 families
Meat - 100 families
Wheat (bread) – 240 families
Vegetables – 300 families

CHAPTER FOURTEEN

SPIRITUAL BEINGS
STAND BEHIND MATTER

In her book *The Field,* Lynne McTaggart summarizes much of quantum physics research and states that "the zero point field [is] an ocean of microscopic vibrations in the space between things...the very underpinning of our universe is a heaving sea of energy—one vast quantum field. If this is true, everything would be connected to everything else like some invisible web."[2]

To help the reader understand biodynamic farming better, I have to share some ideas based on my understanding of how the world works. Lynne McTaggart's comments might help make sense and give some credibility to my views, although I go a step further into the metaphysical. Within this quantum field, I believe there is a spiritual world that is beyond time, space, and energy. Just as our Earth has consciousness and we can talk about the being of Gaia, so every planet and star in the universe is a being, and there are billions. Because there are so many levels of the universe, it is difficult to generalize, but through ages of time, it seems our world has condensed from the spiritual world and that, behind all physical matter, there are spiritual

2. McTaggart, *The Field,* xxvii.

beings. For instance, there is a being that forms the essence of nitrogen. Over eons of time, that essence condensed out of the spiritual into the substance of earthly nitrogen. Over time, the spirit withdrew from matter, so that now we experience nitrogen only as the mineralized end of a process. This condensed state is the level at which conventional science works, but there are other levels of existence, which I will describe.

Within the mineral world, forces are still at work. Science tends to think purely in quantitative terms, whereas minerals all have energetic qualitative values as well. For instance, there is a whole field of healing that uses the energy of crystals. Various crystals have qualities, energies that can heal conditions of the soul. We receive certain energies from the ground if we live in a place that sits on granite or limestone. Metals, too, have an influence on the world. Iron has the quality of hardness, which makes it suitable for either weapons or plowshares. A very high concentration of iron is found in the mighty oak. A verse about iron by Hemleban reads:

> The knobly oak tree speaks,
> Servant of the iron Mars,
> O Man, be rooted in the deeps
> And reach up to the heights,
> Be active and strong,
> Be fighter, knight, protector.

We have all these forces and substances within us, which make us citizens of this mineral world. As farmers, we work with those substances and forces. Depending on how we farm, these qualities may be stronger or weaker. It makes a difference whether the iron we get from our vegetables comes from synthetically fertilized soil or from living soil in which forces from the cosmos are active. In deadened soil, it is possible that the mineral substances are no longer imbued with energy; if so, those energies are no longer available to us to provide the ability to be fighter, a knight, and a protector.

Such forces work only in the mineral realm. When we enter the realm of plants and animals, the living realm, all the rules change. There are three levels to the spiritual world that the farmer can experience while working within nature. First, there are life forces. Although the plant or animal anchors this life force, it is one step above the physical; life as such does not rise out of mineral and chemical interactions. Rather, life is a mighty power passed on in a continued stream of generations that we can experience, for example at birth. When a calf is born and it lunges to rise off the ground for its first drink, the mother patiently stands and nudges the calf into position. Likewise, a seed breaks through the crust of soil to reach the Sun. These let us experience the power of life. Beethoven catches this life force most perfectly in the Ninth Symphony. In "Ode to Joy," the choir sings:

> All creatures drink of joy
> At nature's breast.
> Just and unjust
> Alike taste of her gift;
> She gave us kisses and the fruit of the vine,
> A tried friend to the end.
> Even the worm can feel contentment,
> And the cherub stands before God!

Life pervades the whole cosmos and it is a celebration and hymn to God. The biodynamic preparations work directly to strengthen or harmonize this energy level. Rudolf Steiner terms it the etheric field.

The next level we can experience is the feeling or emotional realm. As we move up the order of animals, these feelings become more pronounced and refined. In addition, the internal organs of our body such as heart, lung, kidney and liver assume functions important to the organization of the body. Traditionally each organ is connected to a different planet and considered the seat of our feelings. For instance, we talk about a lily livered person or somebody having a strong heart. For animals, this life of feeling is internalized, whereas the environment surrounds the plant, which experiences this externally. Steiner named this region of flowing feelings the astral world.

We still have one more spiritual realm, and this shines in from the periphery for both plants and animals. The archetypes of the species are held in the realm of the constellations of the zodiac. When a seed is placed in the ground, it grows into its own kind because it fills in with matter the formative forces coming from the zodiac. For humans it is a little different, as the spirit of each individual person is like a whole species of the animal kingdom. Our being has its home in the starry heavens.

Because people with spiritual vision have developed different organs of perception, each will see and experience different aspects of the spiritual world. After all these years of meditation, I can now see the beginnings of the etheric body of a plant, a beautiful, transparent blue cloud-like shape surrounding the plant. As a beginner with this vision, it takes time to see it. I look at the plants and bring up a meditation, which can vary. The core meditation for me is to be grateful that the plant is so faithful to her task in life and that every day she makes it possible for the world to survive.

In Ecuador, the capital of the flower world, we have a beautiful flower garden, and one particular white rose gives off a wonderful perfume. I am able to steep my senses in the sounds of nature, the delicate smells from the garden and earth, and to take in the breathtaking view over the valley and mountains. I can start to experience the spiritual world by practicing thinking with feeling. It is different from ordinary observation or intellectual thinking in that I actively, and with as much will as I can muster think in great detail about the flower and then fill this thinking with feelings of love and gratitude. Often after doing this for a while, a blue cloud formation starts to appear around the flower

CHAPTER FIFTEEN

THE FARM AS AN ECOSYSTEM

How can we experience the farm ecosystem on the four levels just described? As a farmer, one tries to be conscious of these aspects. Living beings depend on the vitality that streams to the Earth from the Sun. In our modern consciousness, we think of sunlight only as energy. But there is a life element to it as well.

In Eastern philosophy, one talks about *prana,* or life force. The Sun does deliver solar energy, and from this view we can think of a farm as a huge solar collector. Plants use about 1.5% of the sunlight that reaches them, and this is probably a good thing, since the soil would be unable to supply the needed nutrients for a greater growth rate. Human beings receive some of the energy stored in the plants by eating vegetables, fruits, and grains, but it is hard to maintain the fertility of the farm when we produce food only for human consumption. People do not return much fertility to the land, but cows do because they are ruminants. Cows love grass and legumes, and so they excel at converting the stored energy in the grass into food and renewing soil fertility. They produce meat and milk along with tons of manure and urine, which are a great source of fertility.

It is amazing how compost made with cow manure improves both the fertility and the organic makeup of the soil. Hay and pasture are also restorative crops, because they build up fertility and organic matter through the accumulation of root mass, which becomes integrated into the soil. Hay and pasture can also be worked easily into a rotation. With all this fertility, we can grow food for people, such as vegetables, corn, beans, and small grains. This is the point at which we begin to have a farm that is self-sufficient, both in fertility and in the food required to feed the cows.

Within the farm boundary, mentioned earlier, the life forces of the plants are recycled. When the crops grow (grass, for example), they are full of life energy that the cow digests and, in her ruminant stomach, adds her own sentient or astral, or soul, energies. The resulting manure is a life-filled product that can be composted. In the compost pile, we not only transform organic matter into humus, but also retain the life energy that, when spread back on the ground, enlivens the soil.

In the world of nature, there are hosts of nature spirits that we cannot see with our physical eyes because they have no physical body. For example, there are earth, water, air, and fire beings; tree and landscape beings; cloud beings; house beings; and innumerable others. In the past, many people were able to see them, and many indigenous peoples still acknowledge and honor them as part of their lives. These beings are still around us. They carry the consciousness of the farm, and they work on an energetic level. However, it can be difficult for them to assimilate energies that come from outside the farm. Especially hurtful are forces from synthetic fertilizers and herbicides, which they have to assimilate into the farm organism. Part of my practice as a farmer is to honor those beings, and for many years Susan and I would take time each morning to communicate our intended activities for day, explaining our needs and celebrating with them. In this way, we felt that we were able to co-create with them, even if it was at a novice level. We often had the sense that we were heard and helped in our efforts.

The soil on a biodynamic farm has sentient life. Through the use of the biodynamic preparations, we can strengthen and balance the cosmic forces working within the soil. All the forces of the planets and the stars of the constellations are present, ready to help the plant materialize. If you take a walk around a biodynamic farm, you would notice the softness of your steps. If you feel and smell a handful of soil, you notice that it is filled with life. A farmer would say it is a good soil. Such fertile soil is imbued with life forces that are ready to become plants. Into this soil, we place a seed, and all the available etheric and astral forces become engaged in its healthy development.

CHAPTER SIXTEEN

THE BIODYNAMIC PREPARATIONS

For many people, biodynamic preparations are incomprehensible. When I first encountered them, I did not question whether they worked, though I have now studied them in depth. I simply accepted them as remedies that strengthen the life forces of the land. Maybe I felt so comfortable because I had gone to a Waldorf school for a few years and had such a good experience there. Or perhaps it was because I was already familiar with Steiner's ideas. Now I feel that they allow me to co-create with the spiritual world. Whatever the case, I want very much to explain their potency to others so that the farms and gardens of others will thrive.

First, are the compost preparations, which are made from plants and whose energies are enhanced by placing them in animal sheaths and then burying them. All this may sound a bit foreign, but their effectiveness has been validated so well over the years that many consider their effectiveness a proven fact.

Next, I will describe the plants and animal sheaths of the preparations and what we do with them.

- The flowers of the yarrow plant are stuffed into a stag's bladder, which is then hung in the summer Sun and buried for the winter.

- Chamomile flowers are made into sausages from the intestines of the cow and buried in the ground for the winter.

- Stinging nettle is compacted into a bunch and placed in the ground for a whole year, starting in the fall.

- Oak bark is ground to a fine consistency and placed in the brain cavity of a cow skull and then set in a wet place such as a stream for the winter.

- Dandelion flowers are stuffed into the mesentery (abdominal cavity lining) of a cow and placed in the ground for the winter.

- Valerian flowers are pressed for their juice which can then be stored in a bottle.

These preparations, except the valerian juice, are placed in the compost pile in separate holes after the pile is completed. I like to put the stinging nettle in the middle, the oak bark and chamomile opposite each other at one end and the yarrow and dandelion opposite each other at the other end. Only a teaspoon of each preparation is needed in each hole. The valerian juice is diluted, stirred, and sprinkled on the pile with a watering can.

There are two other preparations that are used as sprays for the fields. One is made from cow manure, which is put into cow horns and buried for the winter. The other is made from quartz crystals that are ground very fine and then put into a cow horn and laid in the ground for the summer. When used, both of these preparations are diluted with water, stirred for one hour, and then sprayed, one on the ground and one on the plants.

Over the years, my relationship to the preparations has deepened, but it is not a very intellectual connection. My conviction regarding their efficacy grows through experience.

One can purchase these preparations, but tend to be expensive, so there are regional groups that make them. Michaelmas, which falls around the autumn equinox, is the time to make most of the preparations. In East Troy, Wisconsin, Dick and Ruth Zinniker host a gathering each year to make them. They have the oldest working biodynamic farm in the US, started in the forties and now run by the third generation. They have a lovely old stanchion barn that is set up to make the preparations. Ruth gathers and prepares all the ingredients beforehand and everything is ready to go. Usually about thirty people turn up to help, and it becomes a special fall festival.

I needed about 400 cow horns for my farm, so I would bring my own horns and cow manure. About twenty of us at a time could sit on straw bales and fill horns with spoons. Conversation was good, since many of the people knew one another, though met only once a year, so it was a time to catch up.

In another part of the barn, the flowers and sheaths would be waiting to be worked on. When making the preparations, they do not seem strange or esoteric. The experience is closer to alchemy, working with plant and animal energies that are then given over to the earth to be strengthened and transformed. It all seems quite normal and real...futuristic, rather than old fashioned. After burying the preparations, there would be a potluck dinner, a bonfire, and music, deeply enjoyed by all.

To make sense of the preparations, I have made my own relationship to them. I am not much of a chemist, so looking at them from that point of view has not been my focus, although this is certainly possible. Studying the planets and the energies that radiate from them has given me a door through which to relate to them. Many of these ideas come from study material that Bernard Lievegoed gave to farmers in 1951 and, of course, Rudolf Steiner's 1924 *Agriculture Course*. I have accepted certain statements and built on them. I will try to describe these, though more as images than scientifically.

Again, when working with the preparations, a completely new mind-set is needed. Conventional agrarian science holds the view that for every pound of nutrients you take from the soil, you have to find a way to replace it. In biodynamics, one holds the alchemical view that transmutation of the substance can take place, and that potentizing a substance produces an enhanced effect from the energy released.

Potentizing involves taking one part of an extract, diluting it with nine parts water and shaking it for one minute. You then take one part of this solution, add nine parts of water and shake it for one minute. This would be called "D2 potency." In homeopathy, the same medicine can have varying effects depending on the potency. Eugene Kolisko, a pioneer of homeopathic research, performed experiments showing that the number of times a substance is potentized has an effect, even as much as D60 potency. Although conventional scientific tests can no longer detect the original substance, the energy, or blueprint, of that substance is still present in its effect.

Transmutation of substances, when one mineral turns into another, does take place in the living realm, in plants and animals. When the soil is alive and working well, substances can transmute. For instance, the mineral potash or the mineral lime can transform into nitrogen.

As mentioned, behind matter stands spirit, but for spirit to manifest materially it needs something to anchor it. The preparations work as that anchor. They work medicinally so that the plant can attract the substances that it needs for growth and can balance the etheric and astral in the right way to make the plant healthy.

CHAPTER SEVENTEEN

THE HORN MANURE AND HORN SILICA PREPARATIONS

As a farmer, horn manure and horn silica preparations are the two preparations that I have worked with most. They are relatively easy to make and most years I could cover the whole farm two or three times with my spray rig.

The Horn Manure

As described earlier, the horn manure preparation is made by stuffing cow manure into a cow horn and burying it for the winter. When you dig up the horns in the spring, the manure is completely transformed and has lost its manure smell and consistency. Four hundred horns take up a lot of space, so I would use the bucket of my skid loader to dig a hole about two feet deep, eight feet long and six feet wide. Into this I would layer my horns, with earth between each layer.

When the horns are dug up in the spring, the preparation can be stored in clay pots that are then placed in a special box lined with peat moss. The peat moss stops the energy, or forces, from dissipating into the surroundings. When using the preparation, one needs about a handful per acre, or less

Cow horn preparations

when spraying a large area. The preparation is then stirred into the water for one hour.

Stirring is important, because it transfers the imprint, or information, of the preparation into the water. Stirring by hand, say in a five-gallon bucket, requires a nice straight stick about eighteen inches long, which is used to create a vortex in the water through vigorous stirring. Once you have created a funnel vortex in the water that nearly touches the bottom of the bucket, you stir it in the opposite direction. Initially, this creates chaos, but continue stirring until a new vortex is achieved. Stir back and forth for an hour, and then one can spray the desired area with a whiskbroom or backpack sprayer.

I had more than 500 acres to cover, so I used a specially made stirring machine with an electric motor, which stirred about ninety gallons at a time. I would then transfer this to a spray machine that was attached to my tractor, with which I could cover about thirty acres. Usually we could do two loads in an afternoon. Much of my land was either in hay or in pasture, and I would try to spray these fields in the spring and fall and, whenever possible, after making hay. The cultivated ground, where we were going to plant annual crops, such as corn and grain, would be sprayed before planting.

About this preparation, Steiner says,

> By burying the horn with its filling of manure, we preserve in the horn the forces it was accustomed to exert within the cow itself, namely the property of raying back whatever is life giving and astral.... Thus, in the content of the horn, we get a highly concentrated, life-giving manuring force.[1]

The cow is a ruminant with four stomachs that can hold more than fifty gallons of digestive juices. The plants that she consumes are permeated with life forces, to which she adds her own sentient forces, making the manure a very lively substance. When you look at a cow, you can see that she is a very inward, dreamy being. She reflects the whole cosmos in her digestion, and that energy is retained by the manure. All living beings have energies that flow in and out and keep them connected to their environment. The horns and hoofs, which are made from layers of skin, radiate all the cow's forces of digestion back into her stomach. Thus, when you watch a cow eating or chewing her cud, you experience this total absorption that she has in her digestion. Even after you take the horns from a dead cow, they retain their function of radiating the cosmic forces into the manure stuffed into the horns. When you use this preparation on the bare ground before planting or on hay fields and pasture, you stimulate the forces of germination, root development, and growth.

---

1. Steiner, *Agriculture Course*, p. 74.

The Horn silica Preparation

This preparation also uses the cow horn, but instead of using the manure we use silica from quartz crystal. The crystals are finely ground, water is added to make a paste and then the paste stuffed into a horn. The horns are then placed in the ground for the summer and dug up in late fall.

A silica preparation is sprayed early in the morning, preferably soon after sunrise. This made it difficult for us to use this preparation, since we started milking at five o'clock in the morning. Usually, this meant giving up my mornings to sleep in, but it wasn't all bad. I could set everything up the evening before, such as filling the stirring tank with water. Then at five in the morning, I'd flip the switch to the stirring machine while I enjoyed a hour of tea and watching the Sun come up.

Spraying is a relatively simple operation, so driving through the fields and watching the world wake up was enjoyable work. The silica spray complements the horn manure preparation. In the human being, silica is found in the skin and other sensory organs such as the eyes. It is a carrier of the light and formative forces; it helps to make the plant sensitive to the forces that bring quality and form. Whereas the horn manure helps with reproduction and growth, the horn silica preparation enables the plant to attract the forces that make for good nutrition and high quality.

How Does a Plant Feel in a Biodynamic Soil?

Fertile soil that has been treated with the biodynamic preparations is imbued with life. It is sentient and has a desire to become plant-like. A plant is so close to the earth that there is not a great distinction between the root and the surrounding soil. The seed anchors the spiritual archetype within it. When it is placed into the soil and encounters moisture and the soil's potential to become plant, the plant can than grow in a healthy way. The forces of growth and reproduction are available to it, as well as the forces that produce good nutrition and excellent qualities of fragrance, color, and good taste. When the biodynamic compost is spread on the soil, the soil is enlivened and the planetary forces are more available to the plant. The plant's archetype progresses through the planets, starting from the periphery, down to the earth, where it is anchored by the seed. In this way the plant experiences this whole journey.[2]

If the physical world is a reflection of the spiritual world, then this spiritual world needs a way to manifest, to fill the idea with matter. The plant, in fact all living beings, manifest physically with the help of the planets. I like to think of the archetypes held in the regions of constellations tumbling down to earth through the planets. Through the biodynamic preparations, we can strengthen this process as each preparation is related to a planet.

Valerian is the gateway through which Saturn can bring to Earth the blueprint of the archetype of the species. It is the valerian preparation that imbues the soil with the longing to manifest the archetype of the plant.

The dandelion preparation is connected to Jupiter. Jupiter fills out with matter the archetype or idea of the plant. It allows

2. For more on the preparations see the appendix.

the plant to become sensitive and attracts to itself, out of the surrounding environment, what it needs for its growth. This preparation strengthens the nutritive quality as can be experienced in good taste and aroma.

The stinging nettle preparation is connected to Mars. This preparation further encourages growth into space and the forming of substance, again for good nutrition. It does this by making the soil sensitive, so that it makes available to the plant what it truly needs.

The yarrow preparation is connected to Venus. This preparation enlivens the soil so that the plant can absorb the incarnating forces coming from Saturn, Jupiter and Mars into physical substance. It does this by making the life or etheric body of the plant sensitive, so that it can accept the imprint from the planetary formative forces.

The chamomile preparation is connected to Mercury. This preparation brings everything into fluid movement so that the spiritual can adapt to the physical world. It also strengthens the life or etheric body of the plant so that it does not get overpowered by the spiritual.

Oak bark preparation is connected to the Moon. The Moon influences growth and reproduction. If these forces become too strong then disease can occur. This preparation helps with the further stabilizing and balancing of the etheric and astral bodies so that the plant can be healthy.

The two field sprays, the horn manure and horn silica preparations, help the plant to be balanced between growth and reproduction (coming from the Moon) and good nutritive quality (coming from the Sun). Modern farming has accentuated quantity over quality. It is important that we, as farmers, provide a balance of energies, both of quantity and quality, so that the plant does not have rampant growth at the expense of nutritive value.

Rudolf Steiner gave us a way to work with nature's life forces through the preparations. With these we can heal the Earth and grow healthy food that will nourish us both physically and spiritually. Modern agrarian science is not aware of these energies and has no way to work with them. We have the capacity to consciously enter this realm of life forces but we need to open our spiritual eyes. There is a divide between the physical and spiritual worlds but I believe that for many the bridge between the two is becoming easier to cross. I hope that my story and my work with the land will serve to encourage others to follow their own spiritual path.

EPILOGUE

At the time of this writing, April 2010, life has just changed dramatically. On November 29, 2009, we were attacked brutally in our remote house in the mountains of Ecuador. That evening, a man in army uniform had come to our door asking for our help. We knew this was wrong, since it is customary in Ecuador to remain at the gate and call out or whistle. I quickly closed and locked the door. As I backed away, I saw two other men outside, rising from the shadows. They immediately smashed six large windows surrounding our front door, spraying glass throughout our living room. The thought of running outside went through my head, but I did not know where Susan was. I stood my ground, unarmed, protecting my house and wife and hoping for the best. As they rushed me, they turned off the lights and, in the half-dark, started hitting me hard, up and down my body and legs with long poles. I was terrified because they seemed like professionals, well versed in their routine. They wore black masks and gloves and soon beat me to the ground. Once down, they forced me into the bathroom, whispering *"dinero"* (money) and repeatedly hitting me.

One of the men went in search of Susan, who was hiding in a closet and trying to phone for help. I heard splintering wood as they smashed the door down and Susan crying for help as they hit and beat her. They brought her to the bathroom, pulling her by the hair and dumped her on top of me. Lying on the floor face down, I could see, out of the corner of my eye, that they had a pistol and long carving knife. I was starting to wonder what our fate might be. I felt violated but did not dare show any emotion except submissiveness. I wondered if being shot would hurt. What would it be like having a pistol put to the back of my head? Would I regret my life at the last moment? Would I be able to die in peace? What a way to end my life, snuffed out by a bunch of robbers in my own house.

I could not face thinking about my children. They had already lost their mother and would they forgive me for me being careless about my own life? Then I remembered that Susan and I had had our astrology charts done by an excellent astrologer who cost a bunch. He had said that the stars showed that we would live at least into our eighties. I was betting that our money was well spent and that he was right. Unfortunately, we had taken quite a bit of money out of the bank, as we were getting ready to go to the US. We had to pay our two farmers and gardener not only a month's wage in advance, but also a Christmas bonus, which is the custom here.

When the robbers again yanked me to my feet and demanded money, I had no reluctance in showing them my wallet and desk where we had hid our cash. When I told them I had no more, they again hit me across the back of my neck and head and forced me to the ground. However, they must have believed me; after counting the money, more than $2,000, they proceeded to tie me up, forcing my hands and feet behind my back. Susan was already tied up, and later I found out that they had molested her but did not rape her.

In whispers, one guy told me that they would not hit me again, only tie me up. I did not believe him; I was worried that they would come back and really beat me up to get a confession of more money or just for the brutality of it. I was scared that they would use their boots on me or go for my face, like in the movies. Because they were putting so much effort into tying me, it did not make sense that they would kill me. After a while, they carried me to the bathroom, using my head as a battering ram to knock any furniture out of the way. They dumped me on Susan, double-gagged me, and put blankets over our heads. It took them another twenty minutes to finish going through the house and then it became quiet. They were gone.

By this time, I was starting to hurt horribly. I was lying on my chest with my hands and feet tied behind my back, with Susan, similarly bound, next to me. Later, I found out that some of my ribs were fractured and my left arm was swollen but not broken. I tried moving my fingers and noticed that I could work the ropes. No help would be coming until the next morning, so I had to get us untied. I kept my energy level up by staying focused, mad at the robbers.

I worked on the knots for more than an hour. My legs often went into cramps as I forced my feet up to my fingers. Finally, I got my feet untied and my hands loosened to some extent. This allowed me to scoot over to Susan and, lying on the ground, untie her. She was then able to finish untying me. I could finally let my body relax. The sweat was pouring off me, and my body went into shock. All I could do was lie on the sofa and shake. Susan was in better shape and was able to take over.

Because our cell phones had been taken, she drove down to Vilcabamba and got help. We both ended up in the hospital for three days. I had to receive intravenous pain medications, anti-inflammatory drugs, and antibiotics.

What happened to us was a shock both to foreigners and to Ecuadorians. The robbers must have put out a free lunch sign for Vilcabamba, because during the next three months there were twenty-one more robberies, many of them violent. The good thing that came out of this was that neighborhoods began to organize anti-crime groups. Thus, both foreigners and Ecuadorians got together, and the criminals were foiled and have stopped for the time being. Unfortunately, the police have no investigative skills and were unable to help.

Two weeks after the attack was my sixtieth birthday. We had decided not to return to live in our house, but we did clean it up and had my party there with almost sixty people. It was a great celebration, with roughly equal numbers of foreigners and Ecuadorians, lots of good food, and wonderful musician friends who provided music and dancing. We were celebrating being alive, and it was a way of saying that life goes on.

One week later, we held a Christmas party for the Saraguro dance group and were delighted when they changed it into a birthday party for me. They brought two big cakes, gifts for Susan and me, and performed their dances. For me it was very moving; after two years of chaperoning their class every Saturday afternoon, I was not sure if we had made a connection. From their behavior, I could see that the attack had upset them and that they appreciated my support in their lives through the dance class. I felt we had made friends despite the language barrier.

During the afternoon of the evening attack, we had met with three families who were interested in creating an eco-village. We were not sure if it would be on the property where our house was, but we were sure that we wanted to live in some form of community. The attack reinforced this and showed us that our model of living was not tenable.

When we came to Ecuador, we wanted to experience human *being* and not simply value ourselves for our achievements. For two years, we had been able to live an ideal life. After the attack, we decided to put our house on the market, and the next day we had a purchase offer at our offering price, which we accepted. We knew that we needed to be able to carry within us the peace and joy we had received over the last two years in our meditations. Happiness could not depend on a beautiful view...it was time to become global nomads.

We are still processing the benefits of why we had to experience that shadow side of life. After the attack, it was hard to access all our feelings and process them. We have a very good friend in England, Clare Dakin, who was able to go into our pain and, out of her insight, gave us an invocation to help our healing process. Two quotations from the invocation might help explain how I felt.

Great Spirit,
There is more.
I feel wounded,
I feel physically violated,
Physically and emotionally traumatized,
Troubled by the memory of sudden intrusion,
Of violence and fear,
Of wanton destruction,
Of violation of our personal space,
Our beloved bodies,
Our sacred home.
The shadow side of humanity burst in upon us
And had its way with us

Later the invocation continues...

Great Spirit there is more.
I was personally a target.
My body was targeted.
My beingness was ignored.
I was the target of someone else's rage.
My body was used as a place to throw that rage.
I was disappeared by them.
I was invisible against their rage.

The complete invocation covers the full range of our feelings and, at the end of each verse, asks for healing. I included these lines because they really brought up a core issue for me...that of not being seen. We were of no consequence to those three men. They threw their rage of humanity at us, and we were just pawns in their intent to gain.

We were helped by numerous other friends, with energy healing for our pain and a house to live in while we created a new life for ourselves. This house is surrounded by friendly neighbors and is very secure. I soon felt safe enough to get a good night's sleep.

As part of our spiritual practice, Susan and I have affirmations. After the attack, we created a new set of affirmations, some of which state,

> We manifest our sacred intentions, trusting the universe
> to provide what we need to live fully.
> We are deepening our love as we co-create our destinies.
> We spread our books widely and vibrantly, first as gifts
> and then by word-of-mouth like *Anastasia*

Through our books, talks, and workshops, we help people:

> Experience the spirituality of the Earth
> Learn how to steward the Earth
> Start KINS innovation networks
> Trust their intuition to take risks around their values to find
> their destiny paths...in a way that's a good deal for all

The last two affirmations are Susan's. She, too, has just finished writing a book, *The Trojan Horse of Love,* and we literally finished our books on the same day. Hers teaches how to create innovation networks of kindred spirits while taking risks around your values, told through her life story. Her mission in life was to establish her credibility in high finance and then begin to release love through helping introduce social investing to institutional investors. She feels that real progress is now being made.

Both of us feel that we need to step into new roles as elders. We have a lifetime of experience to share with others, and we want to support each other in our dreams. Our first affirmation ("We manifest our sacred intentions, trusting the universe to provide what we need to live fully") is already manifesting. It has amazed us how many good things have happened to us. For example, we have already started our journey as global nomads.

Right now, we just finished our books staying at Chirije, a small coastal resort near Bahia de Caraquaz that needs help. On either side, stretching for miles, there are no houses. It was home to the Chirije people with 6,000 years of history there, with old pottery shards everywhere. Susan is creating one of her KINS innovation networks to bring people here, with the stated mission that "Chirije empowers the healers of the Earth." I will help with the environmental aspects, although this is a very dry climate and lack of water is a problem. We are doing this on a bartering basis. We stay here as guests of the inspirational Tamariz family who own Chirije, and in return they get help from us to bring Chirije to the world.

Before the attack, we were not living within our budget, but were depleting our savings. We had been trying to find a way to earn income, but everything failed. In Ecuador, it is still hard for me to communicate because of the language barrier, and even when I do cross that hurdle, farmers have no money to pay for a consultant. Soon after creating our new affirmations, we received a phone call asking us to meet with some people who had worked out how to support one another's dreams financially. It really is very simple and is based on the truism that generosity comes back tenfold. This will make it possible for both Susan and I to move fully into the gift economy and not have to charge for our work. I have created a website (growbd.org), therough which I hope we can discuss the spirituality of the Earth, especially from a farmer's point of view. My passion has always been to grow spiritually nourishing food, to farm with respect for the Earth, and to help non-farmers feel connected to the land. People's spiritual connection to food and land is in many cases lost. I know that in the coming years, a path will open for me so that I can share my life's work.

That we have found a way to live our dreams is very heartening; it manifested directly from that commando-like attack. We realize that everything can change tomorrow, but we are trying to live in the present and let our path unfold day by day. For me this is often difficult; I would like to sink my teeth into some project related to farming. I tell myself, first learn Spanish and finish your book. Tomorrow will take care of itself. I know it will. Soon, the first part of our residency visa requirement for Ecuador will be fulfilled, and then we will divide our time between the US and Ecuador. This will allow Susan to be more involved with her dream of helping to green the global economy, and both of us will be able to spend more time with our grown children.

I know that I will be able to help people deepen their relationship to this precious Earth. The whole universe is waiting for us as human beings to take our rightful place as part of the whole rather than simply trying to dominate nature. As humanity moves into the new consciousness paradigm, our relationship to the Earth is changing. We are asked to become co-creators with nature and with the spiritual world. We can do this by learning to work with the life forces of the Earth and by moving into this beautiful and exciting new realm. We begin to enter this world by changing ourselves. To know that spirit is within all matter and that we can open ourselves to this world. We can help heal the Earth by working with the formative etheric energies and the nature spirits. I hope my story will help others to deepen their love of nature and the Earth, which has sustained us through the ages of time.

Recently, I started my own website and blog (growbd.org). Being ignorant about the power of such new ways of communicating, I talked with Meri Walker, a social media expert and now good friend. She has been very patient with me as she explained the ins and outs of navigating the internet and designing my website. I hope it will be a place where people will go to deepen their relationship to nature. Please join me at growbd.org and share your thoughts.

APPENDIX

I have added this appendix for those who might like a more detailed explanation about the preparations. Most of the book has been about my life and personal relationship to biodynamics, whereas the appendix is based on research and therefore has a different style.

Yarrow Preparation

The yarrow plant is very rich in sulfur. In his *Agriculture Course,* Steiner states, "Sulfur is the element in protein that plays the role of mediator between the physical in the world and the omnipresent spirit with its formative power." [1] It is this sulfur process or energy that is strengthened by placing the yarrow flowers in a stag's bladder to be hung in a sunny spot and then buried. Steiner says that by using this preparation "the manure once again becomes able to enliven the soil so that it can absorb the fine doses of silicic acid and lead and so on that comes toward the Earth."

The yarrow plant is closely related to the planet Venus, the planet of copper. During the Middle Ages, yarrow was also called Venus's Eyebrows, which shows that there was an old wisdom about these things.

---

1. Steiner, *Agriculture Course* (Kimberton, PA: Bio-Dynamic Farming and Gardening Association, 1993), p. 45.

The animal sheath for this preparation is made from the stag's bladder. The stag is an animal that is closely connected to its surroundings. The antlers are made of bone, which is usually only found inside the body, covered by flesh and skin. When observing a stag you can see how it is totally in tune with what is going on around it—its antlers are like antennas into the wider cosmic environment.

This preparation reminds me of Botticelli's painting of Venus rising out of the ocean. In the same way matter is able to rise out of spirit. By spreading compost that has been enlivened with this energy, plants are better able to attract elements from the cosmos for healthier nutrition.

Chamomile Preparation

Rudolf Steiner states that, by using this preparation, "You will find that your manure not only has a more stable nitrogen content than other manures, but that it also has the ability to enliven the soil so that plant growth is extraordinarily stimulated. Above all, you will get healthier plants." [2]

Chamomile has well-known healing properties. Chamomile tea will sooth a stomachache and drinking it before going to bed will make for a better sleep. If meat starts to go putrid, you can soak it in chamomile tea and it will be good again. Chamomile likes to strengthen and bring forces into movement. It creates a proper balance between the etheric and astral forces.

We use the intestines of a cow as the sheath because it is through the intestinal wall that the digestive juices are secreted

---

2. Ibid., p. 94.

into the substances moving through the digestive tract. It is where the astral forces of the cow (remember the cow has very strong astral forces) are given over to the manure.

In traditional medicine, the intestines are often connected to the planet Mercury, which has the same tendency of flowing and moving. In Roman times, Mercury was the God of thieves and merchants; both ensured that goods move from one person to the next.

It is beyond the scope of this book to explain everything that Steiner says in his *Agriculture Course,* but he tells us there that the carrier of astrality is nitrogen. Therefore when we place this prep in the compost pile, we create an organ that has the ability to create stable nitrogen and also make a right balance between the etheric and astral forces so that the plants growing in a bio-dynamic garden or farm are healthy.

Stinging Nettle

Yarrow, chamomile and stinging nettle all have sulfur to a high degree and so help spiritual energies be incorporated and assimilated into the compost heap and thereby the soil and plant. Steiner says of the nettle preparation,

> The effect will be to make the manure inwardly sensi-tive and receptive, so that it acts as if it were intelligent and does not allow decomposition to take place in the wrong way or let nitrogen escape or anything like that. This addition not only makes the manure intelligent, it also makes the soil more intelligent, so that it individu-

alizes itself and conforms to the particular plants that you grow in it." [3]

This is possible because stinging nettle has lots of iron in it, which relates it to the planet of iron, Mars. Bernard Lievegoed says that the gesture of Mars is the javelin thrower just as he is about to let go of the javelin. [4] It is very directed and forceful. Persons who are anemic lack iron in their blood. It is the force that allows the spiritual archetype of the plant to incarnate into the world. When you use this preparation, it allows the soil to become intelligent so that it knows what the plant needs. Again, this preparation helps in the health of the plant.

Oak Bark

This preparation, made from the bark of an oak tree and put into the skull of the cow where the brain was, works under the influence of the Moon. It controls growth in the ongoing division of cells, and in reproduction, it controls inheritance, which ensures the continuation of type. If these Moon forces become too strong, if the earth is over-stimulated and growth becomes rampant, as can happen during a wet, warm spring, then we start to have unhealthy plants that are prone to attacks from parasites and fungus. The calcium from the oak bark dampens down the too-strong life forces and balance is restored. Steiner says "It restores order when the etheric body is working too strongly, that is, when the astral cannot gain access to the organic entity.... Then we must use the calcium in the very

---

3. Ibid., p. 96.
4. Lievegoed, *Man on the Threshold*, p. 108.

structure in which we find it in the bark of the oak." [5] This preparation allows the Moon forces of growth and reproduction to unfold in a healthy way.

Dandelion

The dandelion plant is under the influence of Jupiter, which takes hold of the archetypes that Saturn brings to the plant and molds and fills out the form. It fills out the skeleton. The dandelion also has a high content of silica, which attracts substances that provide for good nutritive forces in plants. The flowers of the dandelion are wrapped in the mesentery of a cow. The mesentery is a fine membrane that surrounds the digestive organs of the stomach. It is the mesentery that is sensitive to pain, and it becomes a kind of membrane of consciousness of what's going on in the lower organs of the cow. When this preparation is placed in the ground during the winter, it becomes saturated with the forces of silica. Steiner says of this preparation,

> It will give the soil the ability to attract just as much silicic acid from the atmosphere and the cosmos as is needed by the plants. In this way, the plants will become sensitive to everything at work in the environment and then be able to draw in whatever else they need. [6]

In fact, they become so in tune with their surroundings that they know what is available in the surrounding fields and woods and attract it to themselves.

---

5. Steiner, *Agriculture Course*, p. 97.
6. Ibid., p. 99.

Valerian

It is through the spiritual formative forces of valerian that Saturn works. Saturn is the most distant of the visible planets and is the gateway to the spiritual world. It works like a spiritual sheath and encloses the workings of the cosmos. Further, Steiner says that this preparation "will stimulate the manure to relate in the right way to the substance we call phosphorus."[7] In homeopathic medicine, phosphorus is used to strengthen a a person's spiritual "I." For plants, it is the spiritual archetype that is strengthened by valerian. This preparation needs no animal sheath, nor does it need to be buried in the ground. The juice from the flowers can be extracted and stored until needed. When the manure pile is made, then the valerian is diluted and sprinkled all over the pile to work as a protective covering.

7. Ibid., p.100.

CITED WORKS
AND FURTHER READING

Berrevoets, Erik, *Wisdom of the Bees: Principles for Biodynamic Beekeeping*, Great Barrington, MA: 2010.

Cook, Wendy E., *The Biodynamic Food and Cookbook: Real Nutrition that Doesn't Cost the Earth*, London: Clairview Books, 2006.

——, *Foodwise: Understanding What We Eat and How It Affects Us: The Story of Human Nutrition*, London: Clairview Books, 2003.

Klett, Manfred, *Principles of Biodynamic Spray and Compost Preparations*, Edinburgh: Floris Books, 2006.

Koepf, Herbert, *The Biodynamic Farm: Agriculture in Service of the Earth and Humanity*, Great Barrington, MA: SteinerBooks, 2006.

Lievegoed, Bernard, *Man on the Threshold: The Challenges of Inner Development*, Stroud, UK: Hawthorn Press, 1985.

McTaggart, Lynne, *The Field: The Quest for the Secret Force of the Universe*, New York: Harper, 1987.

Osthaus, Karl-Ernst, *The Biodynamic Farm: Developing a Holistic Organism*, Edinburgh: Floris Books, 2011.

Petherick, Tom, *Biodynamics in Practice: Life on a Community Owned Farm: Impressions of Tablehurst and Plaw Hatch, Sussex, England*, London: Rudolf Steiner Press, 2011.

Philbrick, John and Helen, *Gardening for Health and Nutrition: An Introduction to the Method of Biodynamic Gardening*, Great Barrington, MA: SteinerBooks, 1995.

Santer, Ivor, *Green Fingers and Muddy Boots: A Year in the Garden for Children and Families*, Edinburgh: Floris Books, 2009.

Selg, Peter, *The Agriculture Course, Koberwitz, Whitsun 1924: Rudolf Steiner and the Beginnings of Biodynamics*, London: Temple Lodge, 2010.

Spindler, Hermann, *The Demeter Cookbook: Recipes Based on Biodynamic Ingredients: From the Kitchen of the Lukas Klinik,* London: Temple Lodge, 2008.

Steiner, Rudolf, *Agriculture Course: The Birth of the Biodynamic Method,* London: Rudolf Steiner Press, 2004.

———, *Bees,* Great Barrington, MA: SteinerBooks, 1998.

———, *What Is Biodynamics?: A Way to Heal and Revitalize the Earth,* Great Barrington, MA: SteinerBooks, 2005

Thornton Smith, Richard, *Cosmos, Earth, and Nutrition: The Biodynamic Approach to Agriculture,* London: Rudolf Steiner Press, 2009.

Thun, Maria, *The Biodynamic Year: Increasing Yield, Quality and Flavour: 100 Helpful Tips for the Gardener or Smallholder,* London: Temple Lodge, 2007.

Thun, Maria, and Matthias K. Thun, *The Biodynamic Sowing and Planting Calendar* (Europe), Edinburgh: Floris Books, annual.

———, *The North American Biodynamic Sowing and Planting Calendar,* Edinburgh: Floris Books, annual.

Wright, Hilary, *Biodynamic Gardening: For Health and Taste,* Edinburgh: Floris Books, 2009.

About the Author

WALTER MOORA was born in the jungles of Borneo in 1949 of Dutch parents. By the end of high school, he knew he wanted to be a farmer and began his life path. Realizing that conventional farming fights nature instead of working with her, Walter left New Zealand in 1972 to learn how to work with nature through biodynamic methods. For thirty-five years, he has lived in the US and worked on Camphill community farms, as well as on his own. In 1998, Walter's wife was killed in an automobile accident. After years of deep grieving, he met a neighbor who had suffered similar grief and shared his spiritual outlook. Susan Davis had devoted her life to creating "KINS Innovation Networks" in social investing, organics, solar, micro-enterprise, corporate social responsibility and women's empowerment. Through Susan's Capital Missions Company, the couple began the successful Kindred Spirits Network, or KINS (kindredspiritsnetwork.com), and invited social investors and philanthropists to visit our Nokomis Farm to learn how non-farmers can steward the Earth. In 2007, Walter sold his cattle and machinery, Susan spun off her latest KINS network, and they moved for almost a year to Vilcabamba, Ecuador. Walter currently writes, gives talks and workshops, and shares his life with Susan.

Breinigsville, PA USA
28 February 2011
256498BV00001B/5/P